I0823752

FOOTEPRINT

Charlesbridge
TEEN

FOOTEPRINT

Lindsay H. Metcalf

Published by Charlesbridge Teen, an imprint of Charlesbridge
9 Galen Street, Watertown, MA 02472 • www.charlesbridgeteen.com

Library of Congress Cataloging-in-Publication Data
Names: Metcalf, Lindsay H. author
Title: Footeprint: Eunice Newton Foote at the dawn of climate science and Women's rights / Lindsay H. Metcalf.
Description: Watertown, MA: Charlesbridge Teen, [2026] | Audience: Ages 12 and up | Audience: Grades 4-6 | Summary: "Eunice Newton Foote was ahead of her time and was the first to identify carbon dioxide as a pollutant and cause of climate change—in 1856 when most people preferred that women were seen rather than heard."—Provided by publisher.
Identifiers: LCCN 2024058925 (print) | LCCN 2024058926 (ebook) | ISBN 9781623546335 (hardcover) | ISBN 9781632892805 (ebook)
Subjects: LCSH: Foote, Eunice, 1819-1888—Juvenile literature | Climatologists—United States—Biography—Juvenile literature | Women scientists—United States—Biography—Juvenile literature | Scientists—United States—Biography—Juvenile literature | Feminists—United States—Biography—Juvenile literature | Climatic changes—Juvenile literature | Women's rights—United States—History—19th century—Juvenile literature | LCGFT: Biographies
Classification: LCC QC858.F668 M48 2026 (print) | LCC QC858.F668 (ebook) | DDC 551.6092 [B]—dc23/eng/20250521
LC record available at https://lccn.loc.gov/2024058925
LC ebook record available at https://lccn.loc.gov/2024058926

Printed in Türkiye • IMAK
The authorized representative in the EU for product safety and compliance is eucomply OÜPärnu mnt 139b-14, 11317 Tallinn, Estonia, hello@eucompliancepartner.com, +33757690241
(hc) 10 9 8 7 6 5 4 3 2 1

Illustrations done in pen and ink and photography, digitally composed
Text type set in Cabrito
Edited by Karen Boss
Designed by Diane M. Earley
Production supervised by Nicole Turner

To all the kids who've been told
they ask too many questions.
And to Karen Boss,
who encourages me to ask more.

This book is based closely on true events.
Italicized words represent direct quotations
from primary sources with their
original spelling and grammar.

Foreword

Eunice Newton Foote is perhaps one of the most important "lost" figures of nineteenth-century American history. Her contributions as a woman are unparalleled during her time, and her contributions as a scientist—of any gender—are significant. Correcting the historic record to recognize individuals whom time has eliminated from our memory is one of the most important things that historians and authors can do. Unearthing lost people is perhaps as imperative as finding a lost city—it is a service to the present and future when the body of common knowledge is restored or increased.

Lindsay Metcalf has restored Eunice Newton Foote to our memory in an accessible way. Eunice accomplished things in the middle nineteenth century that anyone would be proud of today, and she did it while starting her life on the frontier and then avidly pursuing education and enlightenment at a time when it was all but forbidden for women to delve deeply into intellectual topics like science.

We are all educators in this world, whether we share our knowledge in a classroom or through the pages of a book or in impassioned conversations. The information contained here is of paramount importance as we aspire to know and understand the world, to understand who came before us, and to understand how our only limitations are the ones we place upon ourselves. Eunice's example is a beacon. Her actions at the first Women's Rights Convention in 1848, her painstaking research in her laboratory, and her dedication to the people in her life are inspirational.

Let your mind embrace this story, so aptly and brilliantly told.

Leif HerrGesell, author/historian

Part One

They fully realized the work before them. Possessed of that material of which martyrs are made, they dispised the counsels of fear,

and worked with untiring zeal,
for the purpose of giving women
every fair chance in life.

—Mary Foote Henderson,
Eunice's eldest daughter,
February 16, 1890

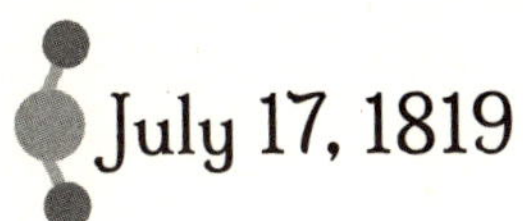

July 17, 1819

Born

The youngest of twelve
yawns,
greets the dawn
of her own revolution.

This puff of ideas is Eunice,
named for her justice-seeking aunt
who learned the Ojibwe, Seneca
& Delaware languages,
who defended her Native neighbors
in an Ohio court when they
were falsely accused of a crime.

Little Eunice is a Newton,
cousin to the very scientist
who plucked the law of gravity
from an apple tree in England—
Sir Isaac Newton.

All around her, coal burns—
ancient life reborn
to heat water into steam,
to power an evolution
of industry.

Justice & science
fold into curious Eunice,
mold into her bones.

She enters the world
in Goshen, Connecticut,
kicking the glass ceiling
to breathe the air above.

1824

Migration

Young Eunice incubates
as her family cracks
their cocoon in Connecticut:

land of mulberry trees—
the legacy of King George's
demand for worms to spin
elegance from their bodies.

Each
silk thread
spans
a cocoon ten yards;
it wraps & wraps,

protecting the body
for the treasure to come,

protecting the body
until it is time

to fly.

Poverty

Life in the silk state
has lost its sheen.

Passed down for generations,
the Newtons' fifty-acre
homestead in Goshen

makes precious little money.
After those fifty acres sell
for a few hundred dollars,

Grandmother Newton
boards a covered wagon
headed to Ohio with Eunice's uncle,

& no thread remains
to tether the family.

East Bloomfield, New York

Eunice blooms on a farm
alongside the flowering meadows
of buttercups, daisies & black raspberries
for which her new town is named.

After the whack of axes,
cropland emerges from the forest
near an ancient road worn
by the Cayuga & Seneca tribes.

As a child—
dirt on her dress,
food in her hair—
Eunice becomes an aunt
to her parents' first grandchild.

Soon the oldest Newton daughter,
Cynthia, has borne a full brood.

Nieces & nephews
function like cousins,
separated from Eunice

by a few years in age,
a few inches in height.

Before Eunice turns ten,
the cousins' branch of the family
breaks off to Michigan Territory,

leaving Eunice to grow up
as the baby once more.

Up

Excelsior! Latin for "Ever Upward," New York's state motto & perhaps Eunice's, too.

1834

Scientist

Until now

man of science

has been the preferred term
for someone who studies
the discipline of natural philosophy.

What happens when
a woman
does that practicing?

Mary Sommerville of Scotland
has delivered such elegant texts
about mathematics,
astronomy,
geology,
chemistry,
physics,
that a new term crowns:

Scientist.

Be it known.

1835

Faith

Under cotton clouds & a birdsong breeze,
the hilltop farm touches heaven.
But as Eunice's father's health declines,
so does the flow
          of plowing,
                    seeding,
          harvesting
& money.

Isaac Minot Newton—
          one of many Isaacs
          in this family tree—
spends his final days & months

alone

in his room
with a bible.

Eunice's father dies
with unwavering faith—
in God, the Newton legacy of science
& his wife & six daughters.

1836

Study

Few in the early 1800s
          believe girls should
          bother with science.

Most in the early 1800s
          believe girls should sew,
                    bake,
                    clean.

Most in the early 1800s
          believe girls should not ask
               why the sky is blue,
               why the air feels warm,
               why they must dress
                    in heat-trapping layers.

Questions
          lead to changes
               lead to revolution.

Eunice's puff of ideas expands,
increases pressure,
seeks

to esc a p e.

Gravity

Miss Emma Willard
believes girls
should bother with science.

Miss Emma Willard
believes girls should ask
why the sky is blue,
why the air feels warm,
why they must dress
in heat-trapping layers.

Miss Emma Willard
has become famous
for building the first school
for women's higher education—

a school that teaches girls
how they can soar.

For decades, she has carved her legacy
brick by brick—
traveling state by state,
country by country.

She lobbied the New York legislature
to trust girls' potential with their money,

not as
satellites of men

but as
primary existences.

Eunice is ready
to leap into that space
& discover the power
of her own gravity.

Test

Hundreds of girls like Eunice
have transplanted to
Emma Willard's famous school,
which serves hearty entrees
of botany alongside ballet,
chemistry alongside Latin.

Many in the 1800s
weaponize biology,
fire cannons of insults
against women,
born "inferior" to men.

Troy Female Seminary
serves girls a well-rounded education—
a revolutionary idea rooted in truth:

their brains are built for knowledge,
for pushing past artificial limits
toward possibilities limited only
by imagination.

Few farmers' daughters can afford
$240 for forty-four weeks
at a far-flung boarding school.

But what science-minded family
could resist the only seminary in New York
with a mission to teach science to girls?

The Newtons scrimp & save
the family legacy by investing
in curious Eunice,
with wings
poised

to

fly.

Progress

At fifteen, Eunice glides
east across New York,
along the Erie Canal,
an engineering *marvel*

forty feet wide,

stolen of life—

nature carved

into a river

few would question

as progress.

Pulled by mules on a towpath,
the canal boat takes only days to traverse
what would take a stagecoach weeks.

Iron gates groan
        open & closed,
                drain & fill
        with changing elevation,
locks unlocking
        Eunice's fascination
                with innovation.

She Discovers

Eunice's new town of Troy is

home
        of the meat-packer "Uncle Sam,"
        famed for feeding US troops
        during the war of 1812;

home
        of glowing iron & fresh textiles;

home
        of the kind of progress

        that belches carbon,

        that burns fossils into fuel,

        that turns skies to s m o k e.

1836-37

Grow

Eunice inhales
every text, every lecture
with azure eyes cast skyward
for any inspiration
that may rain
down.

Kaleidoscope

Seminary students take some courses
at the nearby Rensselaer Institute,
the first-ever school dedicated
to scientific instruction.

Rensselaer not only accepts *ladies*:
it insists they perform
experiments themselves.
Revolutionary!

The first time Eunice's heel clicks
echo off the lab's kaleidoscope of
thermometers,
hygrometers,
barometers,
hydraulic cylinders,
microscopes,

megascopes,
telescopes,
pumps,
lenses,
prisms,
magnets,
sextants,
bellows—

she wants
to leave fingerprints
on them all.

Friends

The seminary sweeps Eunice
into the orbit of her new roommate,
Catherine Cady.

Cate comes from a family
of wealth and influence—
a family who follows the law
& also molds it anew.

At the helm reigns
Cate's father, Daniel—
a former congressman,
an attorney who argued both
alongside and against
Aaron Burr, the third vice president,
best known for the duel
that doomed Alexander Hamilton.

Tragedy trails Cate to boarding school,
burrowing deep in the Cady home
after calling all five Cady sons to heaven
before they could carry on the family name.

Friends II

Eunice learns from Cate
about one of the Cady sisters:
Elizabeth, four years Cate's senior,
who most closely substitutes a son.
With wit & pluck, Libby explores
                    Greek lessons,
                    dinner debates,
                    equestrian excursions.

Before Cate arrives at school,
Libby completes her studies.
But why stop at finishing school?
Why not live radically,
prove what girls can do
with an education?

Because

no matter how many
masculine habits
their father has allowed,
the Cady sisters can
never attend college
& shame the family.

Speak

While the Cady family
follows the rules,
it's Libby who seeks
to make the law just & fair—
advocating for slavery's end
& women's beginning
        to speak freely,
        to own property,
        to VOTE.

Eunice doesn't know Elizabeth well,
but she knows her struggle,
which blankets Eunice
like a threadbare quilt.

She peers through
worn spots in the fabric,
eyes hungry for light.

Panic

Chaos

q
u
a
K
E
s

through the nation.

Banks expand, sell more land
unlocked by the government's
removal of Native peoples.

As banks lend more money, print it faster;
as canals & railroads grow, grow, GROW;
as bound hands build riches for enslavers;
as cotton mops an oversaturated market;

as prices f
a
l
l

& borrowers fail to pay
& lenders seize property,

a tsunami of stress
S
W
A
M
P
S
the Newton farm.

The Farm

Eunice is needed
back in East Bloomfield.

When her father died,
he willed the land & a legacy
of debt to Eunice's sister Amanda—

independent as a firework,
born on the Fourth of July,
a generation after the date
first meant something.

The law of the Patriots' land
decrees that married women
cannot own property—
only their husbands can.

Mandy remains unleashed to any man,
a loophole that lets her hold title.
She calls Eunice home
after the course in Troy is complete

to help rescue the farm from debt
& maintain the land in the Newton name.

Flow

Sister supporting sister,
neighbor helping neighbor—
mending fences,
splitting wood,
plowing fields,
hoisting stones,
planting seeds,
harvesting connection
while sunbeams kiss necks
& sweat drips into rivers
in the steaming heat.

Promise

Under the weight of unpaid loans,
the farm's future hangs in the balance.
Eunice knows a young lawyer who can help.

Visiting the farmstead,
Elisha Foote gathers evidence
to defend the Newton family,

gathers the courage to court
Amanda's younger sister, whose
sky-colored eyes ply Elisha's mind.

Elisha and Eunice share much in common:
he learned the law under Judge Cady;
she lived with Cady's daughter Cate.

He excelled in chemistry & technology
at Albany Academy, not far from Troy where
Eunice learned the art of experimentation.

Perhaps Eunice met him then,
spending holidays under the Cady roof
because her own lay too far away.

This man, a decade older,
sees Eunice for her beauty & her brain.
This promising young lawyer

has the smarts & ambition
to keep pace. With his zeal for invention,
his skill with math & statistics,

his respect for Eunice's love of science,
his respect for Eunice's love,
his respect for Eunice,

the pair vow to leap with both feet
into a revolution
& a future

united.

July 1839

Harvest

Reaper rattles behind a pair of horses,
slicing & flattening amber waves.

Summon all strength
to rake wheat stalks,
bend
& bind them.
Prop scratchy bundles
in floppy shocks,
keeping
heads of
grain aloft.
Let cure in the sun.

Pitch bundles on wagons.

Find a rhythm:
heave, heave.

Horses lead:
clip-clop

to the threshing machine.

Separate grain from chaff,
wheat berries for flour,
straw bedding for livestock,
every bit a gift from the earth.

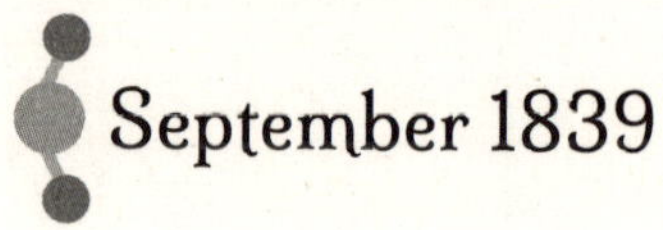

September 1839

Property

Unwelcome news
snakes ~~~~~~
back to Eunice.

The same law that shackled Mandy's hands
& kept her from marrying to keep the farm
has betrayed Miss Emma Willard,

who built a life on her own,
who built a school to mold girls' minds,
who wrote history,
wrote the geography,
education
& poetry books
Eunice devoured as she scoured
the sky for a shimmering future.

Miss Emma Willard, coerced to retire
from Troy Female Seminary
to build a life with her new husband.

Gossip

At fifty-two, Miss Willard,
a widow from her first marriage—
a marriage that helped her grow

into the woman Eunice knows—
has fallen under the charm of another man,
has vowed to become Mrs. again.

She picked up life
& time to write in Boston,
signed away her property
to the Mr. she trusted
with forever. But this man

morphed
into a mean,
controlling
drunk.

And although Emma divorced
herself of his name, the law
decreed that her horse,
carriage,
furniture,
home
remained wedded to him.

Gossip about the celebrity split
splashed across the papers:

She *distinguished, accomplished,*
He *tyrannical, unprincipled,*

came to the conclusion
of losing his rib
rather than her property.

Pressing

Eunice considers the commitment
she is preparing to make to Elisha.

Can she achieve her dreams if she marries?
Can she achieve them if she doesn't?

Eunice folds up
the newspaper,

returns to preserving
the family property

& shoulders
the atmospheric pressure

of unfairness.

1841

It Is Decided

This is the life she wants.
This is the man she trusts.
This is the man she trusts

with her life.

At twenty-two, Eunice is ready.
At thirty-two, so is Elisha.
Add them up & together
their future is greater.

Opportunity follows
rails of steel
          fired with coal—
forging new paths
for trains & people.

Perhaps it's the new tracks
of the Auburn & Rochester Railroad
that whisk these newlyweds
to a future in Seneca Falls,

where Elisha works as district attorney,
where glacial lakes gouge the land
like an eagle's talons.

Eunice opens her wings
& rides the wind
whispering her dreams:

settle down,
but don't settle.

1842

Regulating Heat

Her belly swells
as summer swelters,
heat blooming inside,
due anytime.

Heat sears Elisha's beard.
He fills spare moments
between reading cases,
  arguing in court as district attorney
    & authoring a book about calculus,
      a kind of math invented
      by that famous Newton, Sir Isaac.

All the while, man & wife tinker
side by side in their homebuilt laboratory,
honing the mechanics for changing the world
where trains tow the future past their door.

Before baby comes,
Eunice has a breakthrough.
This is the one,
she assures Elisha.

This is the invention
that will erase our worries
about money.

This one
is worth the price
of time & hope.

May 26, 1842

Regulating Heat II

Eunice's first patented invention,
Regulating the Heat of Stoves,

arranges rods, levers, valves
into a damper that controls
when the heat shall rise
above any required degree—

a damper that clamps down,
cuts off oxygen,
keeps heat *uniform—*
a more perfect control

that can also open,
feed the flame to meet
the requisite intensity

while looking prim
& ornate & beautiful.

US Patent No. 2,636

for a new parlor stove
is dampered by,
is signed by,
is credited to

the man of the house.

Because of the law,
because of the world,
Eunice decides
her invention's best chance
rests in Elisha's hands.

Be it known.

Celebration

Recording a patent
is cause for celebration,

cause for working connections,
cause for convincing someone

to believe in this seed,
to plant it & water it to life.

Elisha takes on that duty,
becomes the face of the invention.

Eunice longs for the day
when her face will be enough.

Her puff of ideas swirls on undissipated,
yet the fog around the effort

of growing life inside her
muffles the brainstorm.

For now.

July 21, 1842

Fire

Childbirth.

A manifestation
of woman as goddess,
warrior inventor,
forging glory
from nothing.

Midwife defeats Death
looming bedside
with his harvest scythe.

Must Not Cry, Eunice is told
          while sweat beads.

Must Not Scream
                    while pain
                              SEARS.

Behave Properly
as the head

e
r u pts.

The Child Has Not Been Told

So the baby breathes
as intended,
oxygen filling
pure lungs

& she

S
C
R
E
A
M
S

carbon awakening
a woman once numb

now

fully

alive.

It’s a Girl

They’ll call her Mary

after the mother of God,

after Eunice’s older sister
            whom Death punctuated
                        five years ago at age twenty-three,
                                    ending her feminine sentence
near the beginning.

They’ll call her.

They’ll remember her name.

Be   it   known.

Part Two

They might become
instead of possible burdens
or indulgent men,

comparatively independent
and useful citizens.

—Mary Foote Henderson,
patent cosigner,
February 16, 1890

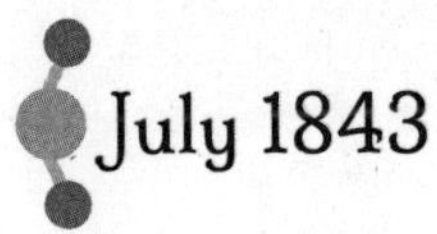

July 1843

Encounter

Sweet baby Mary establishes herself
with the nation's political elite—
her place in the future—
just a week past her first birthday.

A decade and a half after his reign
as the sixth president, John Quincy Adams
passes through the Finger Lakes region
on his summer tour as a US representative,
inspiring a procession of carriages
& military fanfare
& band music
& cannons.

How many hands has this MAN
shaken in his seventy-six years?
How many speeches delivered?
Will he remember the Footes?
Elisha, a district attorney?
Will he remember bouncing Mary
on his knee?

Living History

Mary will hear the stories
& remember this anointment by the MAN
whose father helped draft the
Declaration of Independence. This MAN
who as a boy watched Patriots
& Redcoats
battle near his family's
Massachusetts farm.
This bald & brilliant MAN
who stood alone when it mattered—

tirelessly advocating for slavery's end
& for the creation of a national center
for the increase and diffusion of knowledge
among men,
whatever that means.
A university?
Observatory?
Place of research?
Library?
Publishing house?
Museum?

And the Federalists just might succeed
in their quest to give life to the so-called

Smithsonian Institution

from a scientist's dying gift.

Eunice Wonders

Will this

    Smithsonian Institution

ever be open to women?

Perhaps these thoughts are
interrupted

when Mary fusses,
unimpressed with power

& Eunice rushes to liberate her
from the grip of   MAN

& Eunice lifts her.

        Excelsior!

Real

Eunice's great accomplishment—
her patented, self-regulating stove—
materializes in Seneca Falls,

thanks to a local manufacturer
named Erastus Partridge.
Her concept works. Truly!

Early reviews glow
like the belly of a stove:
Well known
as the most fuel saving
as well as the safest
and most pleasant stove
now in use.

Influence

In man's world,
Elisha exits his role
as district attorney

& grips the ladder's next rung
as judge of the Seneca County
Court of the Common Pleas,

where—please & thank you—
this good man of influence
can wield his influence for good.

1844

Oven

Eunice's belly blooms.
Again.

She sheds her corset
again

to allow new life
to become
who it will be.

A boy
destined
for the world?

Or another girl

who must fight

for her humanity?

October 24, 1844

Deliverance

They call it a sickness—

this condition of delivering
new life from one's womb.

Eunice has been here before—
fire in her belly,
heat expelling.

She is the sun—
visualizing health,
manifesting it
    for this child,
        for herself,
            for women
                & for the world.

In a haze,
baby girl number two
                              enters
                              this man's world.

She'll be Augusta,
                              or Gus, Guss, or Gussy,

whomever she wants—
Eunice will make sure of that.

She'll be a loved,
valued,
powerful

                              WOMAN.

Months

Nurturing,
nursing,
pacing,
playing,
rocking,
shushing

grounds Eunice in the nest.

Stiff as her wings have grown,
they remain unclipped & ready

for opportunity.

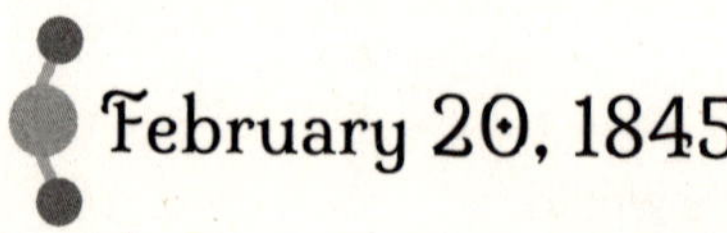

February 20, 1845

Her Time

Behold!

The New York legislature
passes an act that allows
a married woman to claim

a patent for her own invention—

to own & to hold,
from this day forward,

all the credit, all the earnings

as if she were unmarried.

Almost three years
after the stove patent berths,
lawmakers have cleared the way
for Eunice to birth a patent all her own—

a manifestation
of woman as goddess,
warrior inventor,
forging glory
from nothing.

Mother of Two

Eunice has two girls now.

Mary at two and a half,
all wriggly legs running.

And sweet Augusta,
four months young,
cooing, drooling, snoozing
long enough to grant Eunice
the gift of dreams.

These last three years, a fog
of babes rocked by moonlight,
diapers changed,
noses wiped,

first steps cheered—
both the babies'

& Elisha's,

into the world of invention.

Eunice laces her boots
& readies herself
to take her own

first

steps.

Investment

While Eunice finds her footing,
maps her path, an opportunity arises

for Elisha.

A local landowner, bankrupt
after the Panic of 1837,
sends the Seneca Falls fixer-upper
he built for his bride—
sunny porch,
orchard
& river view—
to auction.

Elisha & an old law friend
swoop in with a bid.

One man's panic,
another man's treasure.

Flip

The investors don't plan
to live in this house—
they will buy cheap,
    flip the property,
        make money.

Elisha's partner is Edward Bayard,
who also lives & works in Seneca Falls
& who married the oldest Cady sister,
Tryphena.

Soon these investors have a nibble:
Bayard's father-in-law & Elisha's mentor,

Daniel Cady,

    who has risen to associate justice
    of the New York State Supreme Court,

    who has risen to power
    where his conservative ideas

    & his view of the Constitution
    rule supreme,

    who spends $2,500 to rule this fixer-upper
    for another daughter—

Elizabeth.

1846

Light

Elisha resigns as judge
so he can practice law again.
Eunice practices the art
of juggling motherhood
with her dreams.

Elsewhere,
humans tinker
& discover
another new fuel
derived from fossils.

Kerosene burns bright,
lights up a room.

This clear liquid, distilled
from seams of ashy coal,
is touted as the future—

a future when whales
are no longer killed,
reduced to oil
to burn for light.

The Footes huddle in their home lab
on North Park Street in Seneca Falls,

home of pump houses,
home of a waterway to the Erie Canal,
which links people,
things
& progressive ideas,
such as women's rights
& abolishing slavery.

Eunice & Elisha tinker
& discover
together.

Their love of science,
equality
& each other

burns bright,
lights up a room.

One of these forms of light

is more sustainable

than the other.

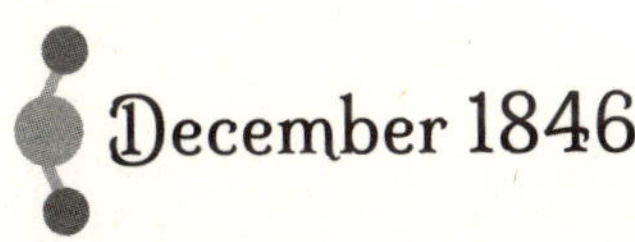

December 1846

Smithsonian Institution

It's been announced
that Professor Joseph Henry

from the College of New Jersey
will lead the new Smithsonian.

Science's rising star
is not only familiar to the Footes,
he's an old friend of Elisha's.

Eunice traces Henry's comet,
unsure of where it will lead
but certain it will blaze a trail.

Joseph Henry

Elisha's history with Professor Henry
stretches to their early lives in Albany.

A young Henry began teaching
mathematics & natural philosophy

at Albany Academy the year
before Elisha enrolled in 1827.

In that red sandstone building,
steps from the state capitol,

Henry debuted his discovery
about electromagnetism:

a small battery could power lift
a shocking 750 pounds.

From then on, the connection
between Elisha & the professor

also proved
magnetic.

May 1847

Radical

Catherine Cady's sister Libby—
not Mrs. Henry B. Stanton,
not Elizabeth Stanton, but
Elizabeth Cady Stanton, thank you.

She and her family look
to move where their lungs
& their anti-slavery agenda
can breathe.

The run-down property Elizabeth's father
purchased from Elisha Foote should keep
Elizabeth out of any radical trouble.

The Stantons abandon
Boston's lung-clogging smog
for the clear upstate air of Seneca Falls.

She totes three tornadoes in trousers,
boys

ages five,
three,
almost two,

whom she might call the next generation
of patriarchy.

The Stanton family swarms
the abandoned decade-old house
on Locust Hill. At 32 Washington Street,

they can drink in views of the Seneca River,
harvest fruit from the backyard orchard
& reap the talents of carpenters, painters

& gardeners to transform the property
into a rural refuge. Elizabeth's feisty ideas
have drawn Eunice in since they first met.

March 1848

Catching Wind

Two days after the Smithsonian board
commits $1,000 to the study of weather,
Elisha rides the changing winds
& dashes off a letter to Joseph Henry.

He is quick to announce his invention:
a self-registering apparatus that can record
wind speed, direction & temperature.

A series of iron rods wrapped
in sheet zinc all lead to a pencil
poised above paper, ready to record,

which perhaps may be
useful for the purposes
of the Smithsonian Institution.

Sunny Skies

An advertisement for the Foote
Self Regulating Stove boasts:

500 gentlemen cheerfully
recommend it to the public.

Eunice radiates after this sunny review,
knowing the stove has been judged

not by its feminine creator,
but by its hot concept.

April 1848

Married Women's Property Act

Women in New York
gain a shred of independence
with the state legislature's stamp
on the Married Women's Property Act,

allowing a woman tied to   man
by name and by law
to own property

as if she were a single female

& to stop
being property
of her husband.

Excelsior indeed!
New York lays pavers
for other states to pursue this path.

So where is the press?
Where are the parades?

Women like Elizabeth Cady Stanton
have stalked the halls of the New York
Assembly and Senate since 1836
pushing this legislation.

Libby lobbied & launched petitions
after she struck the word *obey*
from her wedding vows.
Even after she first swelled with child.

Now, finally—
finally
white   men
elected by white   men
have tossed this shred of independence
to their ~~chattel~~ ~~property~~ women.

But nothing is final
& finally
the Libbys & the Eunices
& the beskirted masses
rise with this climatic shift
in rights for women.

Scale

First patents,
now property.

What other rights
could women carve
from the mountain?

Possible

Eunice is happily married to a   man
who values her intellect
& respects her personhood.

But the very notion
that the law supports her
lifts an anvil from her chest.

Her mind soars,

            skims the countryside,

skips across the Finger Lakes

back to East Bloomfield,

where her single sister Amanda

has occupied a world

of one or the other—

own the farm

or marry.

Now she can do both if she wants.
The choice is hers, not the state's.

Something clicks inside Eunice
like the teeth of cogs in motion,
a machine sparking to life.

What possibilities await
her generation?
Her daughters'?

Eunice dares to dream
in a way previously
reserved for men.

Three Days Later

The men

of the New York Senate & Assembly
do enact as follows:

e x p a n s i o n
of the Erie Canal.

Carve, carve, carve.
Albany to Oneida Creek feeder.
One hundred and forty-one miles.

Chug, chug, chug.
Oneida Creek feeder to Eastern Wayne County.
Sixty-seven miles.

Spew, spew, spew.
Eastern Wayne County to Buffalo.
One hundred and fifty-six miles.

More, more, more.
The sum of nine hundred and
ten thousand dollars is appropriated

toward the toddlerhood of industry

and the slow death

of nature.

1848

Intersections

What's this?

An advertisement
screams injustice
from the newspaper.

> *JUST RECEIVED*
> *another style Regulator*
> *for sheet-iron Stoves—the best*
> *and handsomest in the market.*

What is this?

A new company in Seneca Falls
has copied the Foote stove design?

WHAT IS THIS?

The Silsby Manufacturing Company
calls their version *Race Regulator*,
patented to a man named Washburn Race.

A patent!

Four years after Eunice.
Even though *patent* means
someone claimed the invention

first.

Next

Compare the two stoves. Such plagiarism!
Silsby tried, but those cast-iron scrolls
& the oval shape with delicate feet

will not outshine Eunice's handsomer version
with roses & reliefs & ornamentation.

What makes the company's owners
think they have any business
in stove regulators?
Horace Silsby, Washburn Race
& another partner, Birdsall Holly,
should have stuck to making
axes & other tools!

The Footes have a mind to give
Silsby's hackneyed design the ax.

But if they want a race,
a race they shall receive.

Elisha summons them
to the starting block—
                    court.

Victory

Elisha blasts off the starting block
& smokes Race in county court,
winning a judgment that says
Foote's self-regulating
stove patent came first.

Let the record show
that Eunice is ready
to collect her winnings
& glide on ahead
to new horizons.

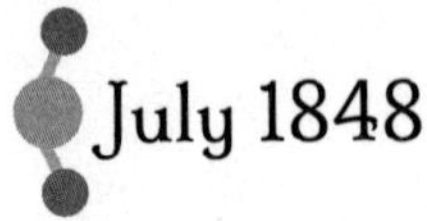

July 1848

Call for Convention

Eunice scans the July 14
Seneca County Courier
seeking some good news—
any good news—

& her eyes land on a call
for a radical convention
five days hence
to discuss the social,
        civil
                and religious condition
and rights of Woman.

The rights of women!

This can only be the work
of Elizabeth Cady Stanton.

Leave it to Libby, new to town
but not to social change.

Eunice knows Libby has attended
abolitionist meetings across New York
& across the Atlantic Ocean.

Ever since men at a world antislavery
gathering in London stuck women
behind a curtain, Libby has sworn
to elevate the cause of women's rights.

So this convention—
this good, good news—
comes as little surprise

& Eunice has been
ready her whole life.

Thoughts

After the dishes
& the washing
& the mending
& the mothering,

Eunice tucks her girls into bed

& perhaps she thinks
of the women before,
of the woman she could be,
of the women yet to be born.

She thinks
of her late aunt Eunice,
who surprised the men around her
by fending off a bear.

She thinks
of her mother, Thirza,
who spent most of her life
pregnant, nursing, mothering
seven girls & five boys.
Thirza has buried four children thus far,
including a baby girl
who never grew up
to be denied her rights.

Eunice thinks
of Mary, days shy of her sixth birthday,
& Augusta, active at three and a half,
both learning the rules—
so many rules
for existing as a woman.

She thinks
of the risks
& rewards
of thinking
this way.

Save the Date

July 17 stands
between today
and the convention.

Eunice's birthday.

How has she passed
twenty-nine years on this planet
with so few legal rights as a citizen?

Eunice vows to spend
her remaining trips around the sun
gifting the promise of freedom

to herself,
to the Aunt Eunices
& the Mother Thirzas,

to the Amandas
carving farmland from forest
& paths of stone,

to the Marys & Gusses
with futures still as fresh
as malleable clay.

She finds July 19
on the calendar
& clears her schedule.

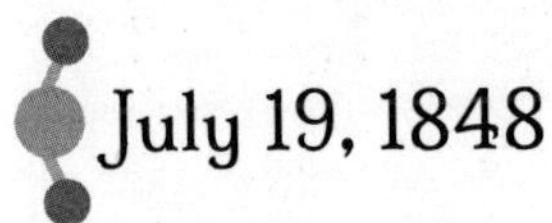

July 19, 1848

Gravity II

Eunice dresses
in the insulating layers
all women are expected to wear
& steps into a steamy morning.

As she walks from her home
on North Park Street,
perhaps she thinks

of the first peoples of this land,
the Cayuga & the Seneca,
forced from their longhouses
without a choice.

Perhaps she thinks
of the freedom seekers
who come under cover of night,
fleeing slavery & violence
to find refuge underground
in this town.

Eunice isn't sure what to expect,
but she is drawn to this women's convention
like a magnet,
or like gravity,

& the closer she draws,
the louder the clatter
of carriages,
of clip-clopping hooves
& the stronger the pull
to the anxious cusp
of change.

Declaration

She's discussed the what-ifs
with Elisha.

What if she's seen?
What if she's reported?
What if merely showing up
damages Elisha's reputation
& their source of income?

But if
Eunice does nothing;
but if
women do nothing;
but if
no one says anything;

she will have to answer
to her girls.

On any scale
the but-ifs outweigh the what-ifs
so strongly that Elisha
makes his judgment
& steps on
to the side that holds his wife
& the only path
toward justice.

Jammed

People stream in
from barges on the canal, trains
& carriages that jam the earthen streets
around the Wesleyan Methodist Chapel.

Of course a women's rights
meeting would take place here,
the five-year-old church
known to allow free speakers
to gather & call for slavery's end.

But this many people?
For women's rights?
During harvest!

The whole village must have
laid down their sickles
& hung up their dish towels!

As the eleven o'clock hour nears
& women crowd around

with daughters,
friends
& some husbands,
the chapel doors remain
locked.

Eunice puffs up
all five-foot-one and three-quarter inches
of herself & cranes
to see.

There!

At the front, Libby Stanton
pushes through, asks her sister
Harriet to help hoist her nephew
through a window
& let the throng inside.

Even at a convention for women's rights,
women must turn to a male for help—
aye, an eleven-year-old boy!—
to unbar doors
& give the signal
to speak.

And Yet

As Eunice & Elisha stack inside
the stuffy redbrick walls of the church,
it's the men who are asked to stay silent.

Three hundred souls pack pews
at the convention to defy convention,
their fervor rising like a fever.

Quakers, Free Soilers, legal reformers.
Speaker after speaker hammers the point:

Resolved,

That woman

is man's equal.

Pen soars across paper
inscribing rights long denied
that women must demand / declare / secure:

The right to OWN PROPERTY,
to KEEP WAGES,
to ATTEND COLLEGE,
to VOTE,

to be anything they want,
to be free,
to be.

Roll Call

Being present is one thing.
Taking attendance is another.

Decision

The next day—
day two of the convention—

Eunice inhales as deeply
as her corset will allow.

Perhaps it's the way she breathes
that changes.

Perhaps the air itself
shifts.

Either way,
Eunice knows
what she must do.

Ink

She files to the front of the line,

inhales the scent of ink,

dips in liquid courage

& tattoos her name to the cause.

Volunteers

Her autograph falls fifth—

Lucretia Mott
Harriet Cady Eaton
Margaret Pryor
Elizabeth Cady Stanton
Eunice Newton Foote

—in a list of sixty-eight women
& thirty-two men whose names
represent a third of the room,
those brave enough to risk

misconception,
misrepresentation,
and ridicule.

Elisha signs, too.

When Eunice's old friend
Elizabeth Cady Stanton calls
for volunteers to float the message
beyond Seneca Falls, Eunice stands—

for women,

for women in science—

& joins the editorial committee.

Currents

The committee & the cause get a boost
from a colossus named Frederick Douglass,
the Black, once-enslaved, abolitionist titan
who shines his light through the *North Star*,
the new Rochester newspaper that declares:

> *Right Is of No Sex.*
> *Truth Is of No Color.*

Eunice puts the final period
on the report of the women's
Declaration of Sentiments,
modeled after the Declaration
of Independence.

She exhales her optimism
toward wider currents,
embracing every part of the country—

> toward *State and national Legislatures*
> toward *the pulpit*
> toward *the press—*

& blows her wishes
like dandelion seeds.

Fall 1848

Science

Word reaches Eunice
that another association has formed,
this one of scientists—
huddling, sharing findings,
enlightening the masses.

Two months after women wake,
seventy-eight men sleep in Philadelphia
& form the American Association
for the Advancement of Science.

Said a man at that first meeting:

The work is not
exclusively for the benefit
of any nation or age.

Yet what of gender?

Eunice knows that
omission
equals oppression.

How can any group
advance
while holding back

any other?

1849

Tracking the Sky

One of the most influential men
at that first meeting of the American
Association for the Advancement of Science

runs the new Smithsonian Institution.
Professor Joseph Henry, the secretary
of the three-year-old hub of science,

is gathering volunteers to play god.
Observers like Elisha from across the US
will report meteorological conditions

to Washington for a revolutionary map
that tracks the rivers of the sky.
Henry plans to assemble the observations

in hopes that someday scholars will use
what they've learned to predict weather.
Imagine! Climates defined!

Farmers could see evidence of which
plants grow best during which months
in which locations. Just imagine!

In the project's first year,
Elisha wrestles numbers
into neat reports on Seneca Falls's

temperature,
humidity,
wind,
clouds,
precipitation.

What Elisha fails to forecast
is how his link to Professor Henry,
the nation's most famous scientist,

will affect his wife's unlikely career
& the trajectory of climate science
& world history.

Passing

While Eunice has been
igniting her power,

her mother's flame flickers.
She has generated all the heat
one life can make.

It's time for her
to leave the earthly form
& soar beyond.

It's time,
whether Eunice is ready
or not.

April 1849

Mother

Here lies
Thirza Root Newton,
who bore twelve children
over three decades,
who buried five of those children.

Here lies Thirza,
born on a bare-branched farm
in December 1773, an early Christmas gift
          to a big clan with Connecticut roots

in wilderness the government paid to tame,
bounties of forty shillings for each
          dead cougar,
hit lists of bears & squirrels & wolves,
          rattlesnakes & blackbirds & jays,

the same acres stripped of trees
to feed furnaces that pour out iron
day & night, belching lost life
          to the sky.

Here lies Thirza,
who moved from Connecticut
          to New York,
          children in tow,
who mourned her husband's death,

who breathed her last at age seventy-six
in East Bloomfield, where roots
feed spring blooms
for generations.

Back

Eunice leaves

East Bloomfield

after burying the Root

of the Newton family,

heads home, praying

spring will show

& something will grow

from sorrow.

Painting

With oils,
Eunice deconstructs

shadows, highlights, ombré tones,
ripples, straights, ridges.

Shade by shade,
stroke by stroke,

she translates each landscape
from three dimensions to two

until the likeness looks true
& Eunice swells with relief.

1849

Wavelengths

The women's wave swells
into
wall
after
wall.

Several states away,
men
in the Tennessee state
legislature claim:

Married women lack independent souls
and thus should not be allowed
to own property.

Men of Tennessee, meet
Amelia Bloomer.

The Seneca Falls
woman wields a wall-
busting sledgehammer:
the first newspaper
for women.

If women have no souls,
then they are free
from obeying any laws, the *Lily* asserts.

Amelia rides the wave
on behalf of all the Eunices
& the Libbys & the Marys
& the Gussys
& the Insert-Your-Names-Here.

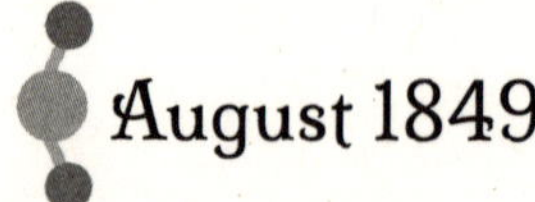

August 1849

Ante-Up

The losers in the lawsuit regarding
the stolen stove-regulator patent—
Silsby et al.—demand
a new trial in circuit court.

They call in the big guns.

One of their attorneys
is former New York governor
& current US senator
William H. Seward.

Elisha readies evidence for battle
& departs for Albany, packing
a trunk with the compressed,
invisible air of Eunice's dreams.

Loop

With Elisha in court,
Eunice has been stuck
at home with the children,

brain space occupied
by active little girls—
thoughts paused,
pushed out by
        tending,
        feeding,
        keeping
them alive.

She prays
for this distraction
to end.

October 1849

Answered

Elisha wins
a $1,500 judgment

in the lawsuit against
Horace Silsby.

Eunice wins
an answer to her prayers:
a priceless period
with Elisha back at home,

creating space for Eunice
to hear her thoughts again
& soar the cloudless blue

of new

beginnings.

1850

Crack

Two years after the boys' club
first shakes hands, two women

c R a C k

the glass ceiling of science:

Maria Mitchell,
Massachusetts astronomer,
& Margaretta Morris,
Pennsylvania entomologist,

the first women elected as members
of the American Association for the
Advancement of Science.

The news buoys Eunice

like a hot-air balloon

rising

according to the laws of physics

& common sense.

Questions

Finally,
science-minded men
reject the lie that biology
weakens women.

Finally
men
of science accept
the possibilities
of women's abilities.

But will they accept Eunice?

What will she say to them?

Which of her ideas should she share?

What climatic shift of consciousness

could she amplify

to the world?

1851

Bloomers

One sort of shift blooms
near Eunice in Seneca Falls:

Women wearing trousers!
Skirts skimming just below the knee!

Heavens, the whispers wrought
by you-know-who. Yes! Libby Stanton

& her cousin & Amelia Bloomer
have chucked their corsets & crinolines.

It started when cousin Elizabeth Smith Miller
tired of fighting fussy fabric in the garden,

a worry she weeded by whacking off
that which tied her down. Adopted

by a fiery reformer like Libby,
amplified by Amelia's newspaper,

the dress draws laughs, stares & tears.
Eunice leaves wardrobe reform to her peers,

instead tackling the gender rules
that shackle scientific minds.

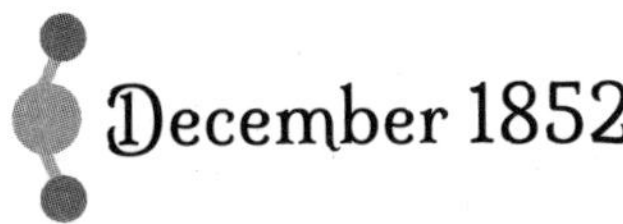

December 1852

Still

Elisha persists in the world of man.
He has defeated Horace Silsby's latest
appeal in circuit court, thanks to the stellar
skills of Henry B. Stanton
(who must debate well
if he wishes to keep up
with his wife, Libby).

But Silsby cues the next round

in the US Supreme Court.

Elisha doesn't bother
hiring a lawyer this time.
He can represent himself against
Silsby's celebrity attorney,
Senator William H. Seward.

A senator!

Who has time
to argue in court
& create laws?

Apparently Not Seward

The defendant:
*Infringe first and look for defenses
afterwards.*

Race's regulator, according to the court:
Not substantially different.

Elisha's case to protect the Foote stove patent:
Affirmed.

Elisha Sues

Elisha has had it
with Horace Silsby's antics.

His stove is still
substantially the same
as the Foote parlor stove.

Silsby must be stopped!
He must pay!

Back to court they go
in a dizzying dance
for power.

Elisha wins.

Horace appeals.

While the men grapple,
Eunice homes in
on discovery.

1856

Since Her Seminary Days

Eunice has kept up on the latest science.

The new issue of *The Scientific American*
debates the effects of the sun's heat on Earth.

A letter from a reader argues that air's density,
rather than the angle of sunrays, explains
why valleys warm more than mountaintops.

This conjecture sends
Eunice's mind soaring

in search of answers.

The Sun

Elisha is working on a hypothesis.

Perhaps he reminds Eunice
that Joseph Henry emphasized
in his recent writings

> *that nearly all the changes*
> *which now take place*
> *at the surface of the earth*
> *are due to the action of the sun.*

Weather. Climate.
Growth. Farms.
Amanda. Women.

Eunice pins her inspiration
& assembles the elements

of her own hypothesis.

Rise

She knows

> that Horace de Saussure
> demonstrated trapped solar heat
> in the 1760s by placing a glass-covered
> hot box atop a mountain & watching
> the mercury rise high enough
> to boil water.

She knows

sunlight pierces the Earth's air envelope,
but some heat cannot escape,
as Frenchman & mathematician
Joseph Fourier described in 1822.

She knows

carbon plied with heat likes to combine
with oxygen; likes to rise, solid to gas;
likes to rise into the air, where it stays.

She knows

scientists have unlocked fixed air
—carbonic acid gas—from limestone
by pouring acid over rock.

And so

when limestone formed, the trapped gas
must have come from the air.

She Knows

Temperatures ran warmer
when limestone formed.

Scholars have discovered that the world
was not always capped in ice, but warm—

carboniferous—with swamps & seas
where now there are none,

tree ferns & tropical plants,
their remnants compressed
                                        into coal & oil,
clues of the Paleozoic past,
                                        fuels of the future.

Some believe the warmth of earlier eras
emanated from Earth's fiery core.

Eunice is not so sure.

Surely

With flashes of
carbonic acid
& limestone
& air density,

questions bubble to the surface.

Could gases

        be the key

                to a warmer

                        Earth?

If so,

which

ones?

Painting II

Painting with facts,
Eunice reconstructs

each invention, each concept,
to tell a science story.

Question by question,
measurement by measurement,

she eliminates possibilities
until only truth remains.

Heat

Eunice inhales inspiration
& gathers tools to trace the story:

pumps,
thermometers,
glass cylinders,

& she waits
. . .
. . .
. . .

for a sunny day,
a shining moment,
when her young sunbeams in dresses
          are occupied.

The Moment

She remembers the cool breathlessness
of a mountaintop & retreats to her lab,

pairing twin glass cylinders,
four inches wide, thirty inches tall,
each cradling a thermometer.

She pumps air
from one vessel   into the other,

& she waits
. . .

. . .

until two columns   of mercury level.
          She sets   the cylinders—
one high pressure,   one low—

in sunlight,

piercing the same blanket of air
that filled the lungs of dinosaurs,

& she waits.

Until

Mercury rises &
condensed air reads hotter by
twenty degrees.

She remembers the thick heat she has felt
by the sea or after a summer shower,

so she tweaks the cylinders,

adding moisture to one, extracting it
from the other

with a bleaching powder called
chloride of lime
—calcium chloride—

& she waits
. . .

so patiently
. . .

while the girls play
or daydream,

oblivious
to the climactic rift
forming in their midst.

Mercury Rises

The humid air heats
twelve more degrees, the spread between
a pleasant stroll & a sweat-soaked slog.

Eunice cannot travel back in time,
cannot mingle with giant lizards,
cannot muddle through their swamps.

But she can
repeat her experiment with different gases.

Hydrogen (the lightest) versus the oxygen
she breathes;
nitrogen (common air) versus the carbonic
acid gas she exhales.

And?
She waits.

Until, Again

The results read clear as glass:
heavier, denser gases grow hotter.

Carbonic acid gas   bests common air
by twenty degrees   &
takes longer to cool.

Eunice dips her pen in a future
even she cannot foresee.

The receiver containing the gas
became itself much heated—
very sensibly more so than the other—
and on being removed,
it was many times as long
in cooling.

In conclusion:

carbonic acid gas
is quick to heat
& slow to cool,

a finding almost as big

as the atmosphere itself.

And So

Eunice waits

. . .

. . .

. . .

to seize an opportunity

to share her spark of discovery

with the world.

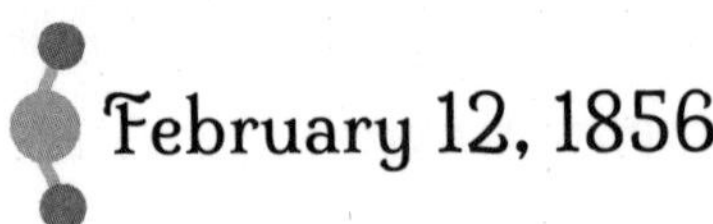

February 12, 1856

Speaking of Sparks

Eunice's nemesis, Horace Silsby,
pivots from putting out
patent-related fires
to putting out actual fires,

fighting fire with fire
when his partner Birdsall Holly
claims Patent No. 14,239
for a rotary steam engine.

Their cast-iron contraption
weighs 9,500 pounds
& clatters across cobblestone
behind a team of horses
like a steamboat on wheels,
coal kindling a boiler.

Steam builds pressure;
pressure turns a turbine;
gears spin faster, faster, faster,
throw water in a constant stream farther,

farther than any standard piston fire engine in the
nascent fire-fighting industry.

Eunice tries to keep
her eyes on her own paper
while accolades roll in
for her competitors.

August 23, 1856

Convention II

Man and wife travel to Albany,
hopeful that their discoveries
will communicate both the cause
of a warmer atmosphere
& the capacity of women.

Sweat beads on Eunice's brow,
trickles down her nape in the August heat.
Elisha gathers his papers
& Eunice follows, heels clicking,
into the state senate chamber.

The American Association
for the Advancement of Science
purports to gather the greatest
scientific minds in the new world.

But with only two exceptions
in its eight years of gatherings,

those minds have all belonged
                    to men.

Elisha presents his work,
On the Heat of the Sun's Rays.

Eunice listens

& she waits for her turn,

next on the program,

centuries from recognition

in her own right.

Today

She is Mrs. Elisha Foote,
shrouding her ideas
in a paper with a title similar
to her husband's:

Circumstances affecting
the Heat of the Sun's Rays.

But she won't present it.
The voice belongs to a man—
a learned, respected man,
but a man nonetheless:

Joseph Henry.

Professor Joseph Henry

This man runs
the Smithsonian Institution,
founded a decade prior.

This friend of Elisha's
& now Eunice's, too,
has so graciously agreed

to put his reputation on the line
& present the paper of a woman,
one *Mrs. Elisha Foote.*

He opines:

Science was of no country
and of no sex. The sphere of woman
embraces not only the beautiful
and the useful, but the true.

The professor reads her conclusion,
her groundbreaking,
earthquaking conclusion
about carbonic acid gas—

An atmosphere of that gas
would give to our earth
a much higher temperature—

yet even tremors do not move this room.

In Conclusion

All her work filters
through a shoddy hypothesis
that women's biological makeup
makes them inferior.

It's No Wonder

At the recommendation
of Professor Henry,
Elisha is voted in as
a member of the association.

Eunice is not.

Still, as Elisha's spouse
she is considered
an unlisted member.

She claims a seat onstage with him
among the who's who of science
at an observatory dedication.

Like the layers of her dress,
she must peel away misbeliefs
one breakthrough at a time.

Reception

Eager to see the write-up
in Tuesday's newspaper,
Eunice flips to the article

about the meeting in Albany.
Toward the bottom,
words steal her breath.

Lastly, two papers were read on the heat
of the sun's rays, by Mr. Elisha Foote, and
Mrs. Elisha, or, as I am requested to call her,
Eunice, Foote.

The latter was read by Professor Henry,
who prefaced it with a few remarks
in reference to the lady. She must be
a charming person if only a quarter
of what the learned chief of the Smithsonian
said be true, graced with every virtue,
and adorned with every accomplishment.

The papers were instructive,
as you may gather from the brief sketch
of that by the lady, but

they would hardly interest your readers.

Painting III

Painting her life,
Eunice brushstrokes her truth
with enough pressure to etch glass

but enough control
to avoid cutting herself

open.

Plucked

Elisha learns his paper
has been chosen for republication
in the *Philosophical Magazine*,

a London periodical known as
the world's premier science journal,
edited by several of the globe's

most famous scientists—
including Professor John Tyndall
of the Royal Society.

The magazine will publish Elisha's
On the Heat in the Sun's Rays,
but not Eunice's work, even though

her conclusions should quake
tsunami-sized waves
across the Atlantic Ocean.

Eunice's emotions spark a storm:

mixing & swirling
heating & cooling
rising
& falling.

Pride swells for Elisha.
Defeat disappoints.
Resolve reignites
like lightning.

1857

Back to the Lab

Twenty-two inches by three inches,
glass tubes fed with a rod dangling
two delicate gold leaves.

Fill with oxygen, hydrogen &
carbonic acid gas—CO_2.

Measure how electricity
races among them.

With an air pump, increase pressure.
Watch the gold leaves repel
each other, then measure
the angle. That's electricity.

Add heat.   Then humidity.
Measure electrical   charges once
and   again.

Hypothesis:

results recur on an atmospheric scale.

Study the writings of white   men:
Becquerel,
Gay-Lussac,
Biot,
Humboldt.
Recognized for brilliant brains.

Discovery thrills Eunice's circuits:
aerial tides,
weather patterns,
climates
are caused by magnetism,
is caused by electricity.

Electrical excitation
~~static~~
may be produced
by condensation
and rarefication of air.

After eight months of experiments,
Eunice inscribes her findings
for the next chapter of herstory.

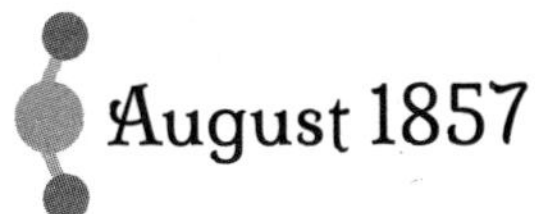

August 1857

Convention III

Behold! It's Eunice.

Recognized
for her brilliant brain,
novel experiment,
meticulous work.
Day three begins—

the annual American Association
for the Advancement of Science
convention in Montreal, where train
after train deposits some two thousand
scientists & their wives, hungry for food
& drink with a dash of knowledge.

Professor Joseph Henry of the Smithsonian
notes the benefits of letting *ladies engage*,
then proceeds to speak on Eunice's behalf.
Henry says he attempted her experiment
On a New Source of Electrical Excitation
& failed, so brava to *Mrs. Elisha Foote.*

Then the man reads his own paper on climate,
with maps marking ocean currents,
wind patterns,
and temperature curves
& nary a footnote
forecasting climatic shifts

in science or society.

But, in November

Eunice is published!

In the *Proceedings of the American Association for the Advancement of Science,*

she writes:

I have ascertained that the compression
or the expansion of atmospheric air
produces an electrical excitation—

meteorologically & metaphorically.

After the male scientific community
ignored,
dismissed,
passed over
Eunice's climate findings,

she returned to work
only to learn that she's the first
woman to grace these pages.

Such joy!
Such accomplishment!
Such anticipation

that finally her news
will flow forth
toward wider currents.

Excelsior.

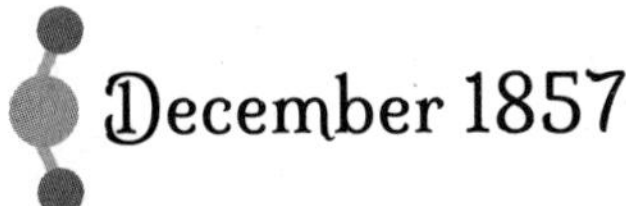

December 1857

Embroiled

Elisha has more pressing matters
than Eunice's new discoveries.

Such is the nature of legal work:
he remains mired in lawsuits.

Yet in the business
of Victorian-era patents,
one must fight for their claim.

Once again
  Elisha climbs the steps
    of the US Capitol Building
      & enters the moody chamber
        of the US Supreme Court
          in his quest to recover
            monies from those
              pernicious people
who
stole
their
invention.

With this case, he hopes
to put this dizzying decade
of fighting for first
to bed.

Stolen

Silsby's lawyers contend
that the Foote patent
is neither original nor useful.

Beneath a dramatic vaulted ceiling,
lamplight amplifying
fiery crimson furnishings,

Elisha stands in the power of truth
& argues otherwise, priming
for a lifetime of fortune.

Thousands of dollars are at stake.
This is the final appeal after years
of meandering through legal battles.

Three years ago, a lower court
ordered Silsby, Race & three other men
in the et al. roster of defendants

to turn over their profits
for self-regulating stoves to Elisha.
They appealed. A jury sided with them.

Now Elisha presents final arguments
to the highest court, sprinting
toward the fortune his family deserves.

Sprinting toward home.

Eunice hopes.

A Divided Supreme Court Decides

The award by the Circuit Court
of damages for an infringement
of the patent
affirmed.

Foote's Self-Regulating Stove wins!
The slate of defendants must pay
damages of $58,500.80—

enough to erase the Footes'
worries about money
for years to come.

Eunice celebrates
her victory
silently.

1858

Partial Victory

After Tyndall's London periodical
snubs Eunice's first paper on gases,
her second paper is picked up,
albeit with notable slights.

The *Philosophical Magazine* removes
Eunice's thoughts about the relationship
between atmospheric electrical excitations
& Earth's magnetism. It also omits her name.

For the purpose of publication,
Eunice is invisible—
merely
Mrs. Elisha Foote.

August 27, 1859

Drake's Folly

A state away in Titusville, Pennsylvania,
black sludge seeps from the soil.

An unemployed rail worker
prospects through shale & doubt,

drilling down to a desperate nothing
for months.

Locals latch the name *Drake's Folly*
to the derrick where Edwin Drake toils,

the geometry of wood where he bores down,
down,

until at 69.5 feet, he stakes a pipe
in the bedrock of more-for-everyone

& oil fountains free—
fills crocks, barrels, washtubs—

for what future no one knows,
except money, money, money,

& oh, perhaps lamp fuel,
a new kind of kerosene

made from black gold instead of coal,
to more easily replace the need

to snuff out whales
for the sake of light.

Whose Story?

As herstory unfolds,
it meshes and molds

with the   history
of industrial
revolution.

The first oil well, bound to burn crude,
to churn carbon skyward, shapes the story

of US

& necessitates Eunice's news
about atmosphere-warming gases.

So which is the greater *folly*?
        (*noun*)
        (A foolish act.)
        (A geyser of a mistake.)
        (A geyser of black gold,
                liquid fossils,
                        oil.)

Is it that no one notices Eunice's discovery

or

that everyone notices Drake's?

Mountains

Now in Saratoga Springs,
a mineral spa town
in the Adirondack foothills,

the family unpacks their trunks
for a long stay at the Union Hall Hotel,
home of grand piazzas & elegant gardens.

From here, Elisha will defend
the famous big-money patent controlling
the manufacture of railroad spikes.

In Saratoga, one can order healing waters
& the most delicious invention—
potato wafers, fried to a crisp.

They say chef George Crum's sister
devised the idea for Saratoga Chips,
but he reaps all the credit.

Eunice tunnels through
that same mountain of patriarchy.

Her pick and shovel:
her home lab back in Seneca Falls.

Her lamp:
her patent-lawyer husband.

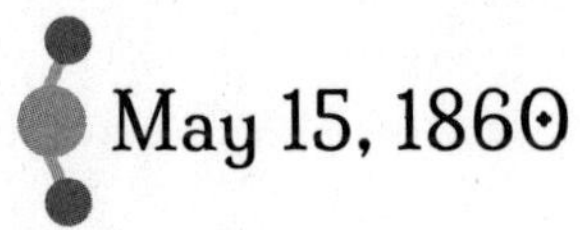

May 15, 1860

Soul

Since 1790, when the US first began
issuing patents, roughly one woman
has earned one each year.

At the top of any mountain,
breath fogs the crisp, rare air.

At the top,
Eunice etches the first patent

in her own name
on the slate of herstory,

for the invention of rubber insoles
to prevent shoes from squeaking.

Patent No. 28,265
bears her signature:
Eunice N. Foote—

a name all her own, cosigned
by the now-grown next generation:
Mary Newton Foote.

Eunice Writes

My improvement relates to the filling.
I make it, in one piece, of vulcanized
india-rubber, pressed by dies

or rollers into its proper shape
for both the shank and the fore part.

It is attached to the inner sole
by rubber cement and then

the bottom is put over it and sewed
or pegged on in the usual manner.

Eunice makes her sole
    watertight,
        cemented,
            durable,
                machined.

One small victory
for squeak-free rubber insoles,

one giant cloud-leap
for woman's soul,
no longer

          cut and pared away
          until reduced
          to its proper shape.

Be it known.

1860

Combustion

Somewhere in France,
a man named Étienne Lenoir

patents the internal combustion engine,
gas-fired instead of steam-powered,

yet still with cylinders,
pistons,
connecting rods,
flywheel.

Somewhere in England,
a man named John Tyndall,

that haughty man who snubbed
Eunice's work for publication,

"discovers" CO_2 as a greenhouse gas
produced by Lenoir's engine,

"discovers" climate science,
a thousand-plus sunrises

after Eunice.

Tyndall

He, a professor
who has all the resources: The Royal
Society! The British Museum! Nobles
who bring him rock salt from abroad!

Eunice, a wife & mother
who has a homebuilt lab,
who works in the gaps of time
when Mary & Augusta don't need her.

He, a man
with luxurious hours,
weeks, days alone
to perfect his methods.

She, a woman.

If

Tyndall lives overseas.
He works different networks;
he has never met Eunice.

He couldn't have stolen her idea,
right?

If

one of Tyndall's experiments appeared
in the same 1856 publication as Eunice's
& he later republished Elisha's work, which
originally landed a page before Eunice's
& Tyndall notoriously disrespects women—

he could have stolen her idea,

right?

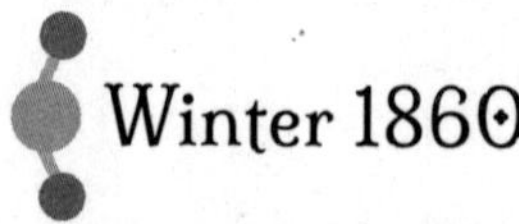

Winter 1860

Combustion II

November 6

Somewhere in the United States
a promising lawyer wins the presidency,
promising to stop the scourge of slavery.

December 20

Somewhere in the United States,
South Carolina breaks off, determined
to keep rivers of sweat & blood running
down Black backs cracked in bondage—

in the name of progress.

April 12, 1861

War

The eleven Southern states that form
the Confederate States of America

attack Fort Sumter in Charleston Harbor,
& after thirty-four hours, Union troops yield.

It's hard to work while fighting rages.
It's hard to invent through upheaval.

It's hard to quibble about stolen ideas
when one's very existence is at stake.

Women's-rights friends pause their cause,
recognizing that no one is free

until all are free.

Women's-rights friends elevate The Cause:

freedom for all.

April 15, 1861

Answering the Call

President Lincoln appeals
to all loyal citizens
to organize & prepare
to fight back.

In Saratoga, Elisha meets
with fellow countrymen
who form the Home Guard
& elect Judge Foote their captain.

With ample pride & patriotism,
without uniforms or guns,
the militia meets twice a week
in a hotel ballroom,

led by a lawyer learning
formations & drills from a book,
which advises him to chalk
out their steps twenty-six inches apart.

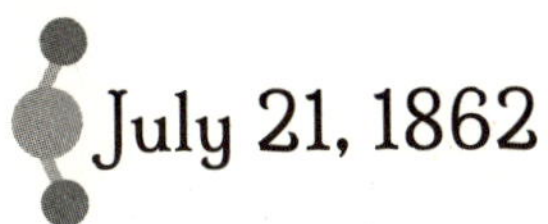

July 21, 1862

Freedom

On Mary's twentieth birthday,
Eunice & Mary apply
for a passport, a privilege
they can afford while
their fellow country-   men

tear one another
                                        apart.

Mary, now a young adult,
is already collecting recipes—
Saratoga potato chips, yes,
but also corn breads like the one
from the US Hotel in Saratoga.

Mary dreams of using her French
& studying the art of French cooking
from Parisian masters.

Eunice dreams of visiting
the world's museums,
seeing the brushstrokes

of world-famous oils
& copying them
as a masterclass
in how to be.

Superficial

The passport examiner sizes up
Eunice's physical traits:

five-foot, one-and-three-quarter-inch frame,
blue-gray eyes,
full forehead,

ordinary nose,
rather large mouth,
small/ordinary chin,

dark-brown hair,
sallow complexion,
oval face.

Anyone overseas
can judge this scientist,
                                inventor,
                        changemaker,
            wife/mother/daughter—
woman—
on the same superficial plane

as this passport examiner—   this man—
from Saratoga Springs.

This

                        is her key

                                                        to the world.

Grand Plans

Eunice & Mary
have made grand plans
with a friend of Eunice's
from Troy Female Seminary.

Mrs. Katherine Youmans—Kate—
& her new husband, the science publisher
Professor Edward Livingston Youmans,

will whisk them to Liverpool, England,
aboard the famous steamship *Great Eastern*,

which, at 691 feet long & 19,000 pounds,
claims to be the largest steamship ever,
with space for 4,000 passengers.

But before her hulking paddle wheels roll—
in five days' time!—
the Footes must secure passports.

Favor

Elisha calls in a favor
from an old colleague,
a friendly foe.

If anyone can muster
the miracle of making
& sending passports quickly
during wartime in Washington,

it's William H. Seward,
who now represents the US
as secretary of state—
the president's closest adviser.

Elisha Writes to Seward

Dear Sir

Mrs Foote & my daughter Mary
are intending to sail for Europe.

They supposed until today that they
would obtain passports in New York,

and I am
very apprehensive
that now
the time is short
to get them
from Washington.

Will you do me the great favor
to send them without delay.

Very Truly Yours
Elisha Foote

The Rest of the Story

Does Seward deliver?
Do the Footes receive their passports?
Do they board the ship in time?

History doesn't record the answer.

Only the sea knows for sure.

Imagine That They Go

If Eunice & Mary receive their passports
& they do proceed to Europe . . .
there will be time for regaling starry tales
while the ship slices across the Atlantic.

In her younger days, Kate sailed on a whaler—
no leviathan like the *Great Eastern*—
from New York around South America's tip
& on to Honolulu to meet her first love,

her childhood sweetheart, Mr. Lee.
They wasted no time reuniting
& exchanged vows before Kate
ever stepped off the ship.

Then Mr. Lee died & left behind
his boatload of fortune, so Kate returned
to New York & embedded in the literary scene.

She has thrown soirees with the biggest names
of the day—Mark Twain, Walt Whitman,
Emerson & Longfellow, to name a few.

But it is the escort for the three ladies
of this would-be traveling party—
Edward Livingston Youmans—
who captured Katherine's heart.

Kate has never borne children.
Her stories are her children.
Oh, the stories she can tell,
of tropical nights
& northern luminaries.

For Eunice, though,
it is surely Edward's tales
that ensnare her attention.

In Particular, the One About Darwin

Edward works for D. Appleton & Co.,
a noted publisher of serious science
based in New York City.

Edward has traversed the Atlantic
many times to connect with
Britain's brightest scientists
& bring their words home.

His latest big acquisition: the US publication
of Charles Darwin's *On the Origin of Species*,
which released stateside a year and a half
before Edward & Kate's wedding
in Saratoga Springs, late 1861.

If Eunice & Mary receive their passports
& they make this voyage to Europe . . .
Eunice surely thrills at the stories
of Darwin's findings on the Galápagos Islands.

So many doubt his theory that all life
evolved from the same ancestor.
For Eunice, it's truly primitive minds
that distrust sound scientific thought.

Hot Air

Perhaps on this would-be voyage
Edward tells of an Irish professor
who has espoused revolutionary
lectures at the Royal Society
of London. Edward hopes
to reprint the lectures of   John Tyndall
for American audiences.

Perhaps a wave of unease swells.
So this is the person Edward
has worked himself up about.

The man

who stole Eunice's credit.

John Tyndall,
the very man
who eschewed
Eunice's breakthrough
about carbonic acid gas
& claimed it for himself.

If Eunice goes on this trip,
does the vast ocean threaten
to swallow her every ambition?

She must not allow it.

Only the Sea Knows

Do Eunice & Mary board
the ship to Europe?
Does Eunice meet Tyndall?
Does she confront him?

The next year, in 1863,
D. Appleton & Co. publishes
one of Tyndall's lectures.

But what
might've
happened
in Europe
stays in Europe.
Only the sea
knows the rest.

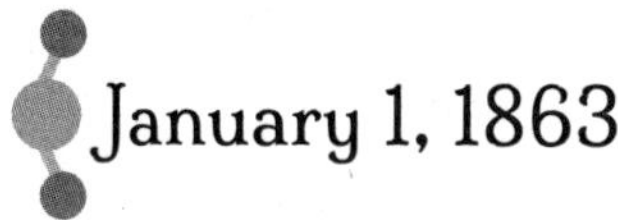

January 1, 1863

Emancipation

Fighting rages. Friends battle.
The world's a whorl of wrath.

Foremost in this moment,
those forced into bondage

are declared free—

emancipation by proclamation,
the lawyer-president's promise

fulfilled.

1864

Applied

Eunice has been chipping away
at the mountain between
women & freedom.

While war rages between North & South,
while soldiers from Saratoga Springs
defend the Union's capital,

Eunice & Elisha send two applications—
two dreams, two futures—
to the Washington patent office.

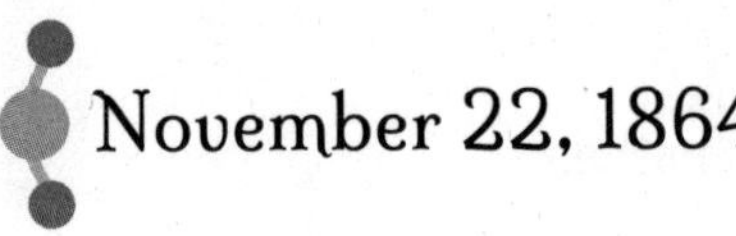

November 22, 1864

Behold

A machine for making
wrapping & printing paper
leaps from Eunice's brain
to correct the creation
of paper deficient in strength,
cheap and inferior.

I have remedied this defect
with a precise array
of cylinders,
slats,
belts
& pulleys

that squeeze the water,
that crosshatch the fibers,
that cement strength by design.

Eunice carves her signature
onto Patent No. 45,149,
itself a pulp of tree fibers,
smashed, smoothed, dried.

Eunice N. Foote,
a name all her own,
witnessed by the woman
who blooms independently
on the family farm, despite
its tangled roots of man—
A. Newton,
her sister Amanda.

Ice

The sisters have arrived in this future,
and they shan't be leaving. And yet,

Mrs. Foote still follows

Elisha.

His patent for a lighter, stronger ice skate,
No. 45,148, slices one digit ahead
of Eunice's patent on the same day.

Be it known.

Work

It's hard to work
when friends are fighting.

It's hard to work
when the country is cleaving.

It's hard to work
when women's work

is not the kind of work
Eunice yearns to do.

Coal

Carbon
Out of place,
Ablaze—
Liberated.

Magic black rock
burns everlasting,
heats homes,
fuels trains & ships,
powers a revolution.

Shovel it in.
Shovel it in.
Burn that black.
Fire those furnaces.
Forge iron & steel.
Pave the map.

Before terrible lizards,
ancient ferns descended swamps
down,
down,
fossilized
over eons,

waiting

for someone
to interrupt.

Interrupter

Word reaches Eunice
that nephew Isaac Newton
(yet another Isaac Newton),
son of her brother Darius,
moved to Cleveland at the start of the war,
made stripping earth his business.

Chop rock,
rip ground,
down
155 feet
under
nature's
chosen
surface.

Over two months,
up to 150

men

built hoists,
laid track,
rigged docks,
shipped out
4,500 tons

of superior coal
each & every
God-given day.

Hundreds of acres took
millions of years to form
hidden treasure,
free until it's gone.

But the men
in charge—Eunice's blood—
have kicked that can
to the next millennium.

1865

Act One

January 31
(inciting incident)

Congress accepts the Thirteenth Amendment
to the Constitution, which abolishes slavery
in the United States, as long as the states unite
and three-fourths of state legislatures agree.

March 1865

Eunice's Babies

Now grown, Mary & Augusta
insist on boarding a steamer
bound for the Confederacy.

Why they must endanger themselves
crossing into shelled-out Charleston
& Savannah, Eunice will never understand.

What if their ship strikes a torpedo?
The papers say the rebels littered
the devastated landscape with bombs
before they evacuated Charleston last month.

The girls will travel in a large party
of Washington's elite—a junket, they call it.
Elisha's cousins from Vermont will be there—
Senator Solomon Foote, whom Mary
will assist, along with his wife, Mary Ann.

Although 19 states have ratified
the Thirteenth Amendment thus far
(even Virginia & Louisiana!),
the roster does not yet include
South Carolina or Georgia.

There will be carriage tours
            & band music

& a gunboat salute
& stars & stripes flying
over Fort Sumter
again.

But the war is not over.
Just in January, rebels arrested Solomon
after he, Union Senate president pro tempore,
visited the Confederate Senate.

Eunice's babies must be safe,
she tells herself.
They must.

Act Two

April 9
(rising action)

Robert E. Lee surrenders
at Appomattox Court House
in Confederate-held Virginia.

April 14
(midpoint/twist)

Abraham Lincoln attends his last play,
receives an actor's life-stealing bullet

while another Confederate zealot breaks
into the home of William H. Seward—

secretary of state,
former New York governor
& the attorney who once
faced off against Elisha Foote—

& that zealot /
Confederate sympathizer /
white supremacist

stabs Seward & six others

who all survive.

April 14, 1865

Crisis

What does it mean
when the one doing the freeing
is shot dead—
imprisoned in time?

What does it mean
when the comedy onstage becomes
a tragedy in the wings?

This tragicomedy—
this all-is-lost moment
in the hero's journey—

is not Shakespeare

but a cliché of Greek mythology,

at once an undoing

& an ignition.

Act Three

May 10
(climax)

Confederate President
Jefferson Davis
is captured! War ends!

June 19
(falling action)

The last of the enslaved people
learn about the Emancipation Proclamation,
claim their freedom two-and-a-half years

late.

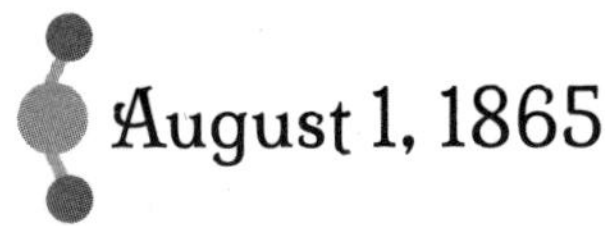

August 1, 1865

Interlude

On his fifty-sixth birthday, Judge Elisha
ascends to national influence
in the US Patent Office.
A gentleman
of the highest ability
and character,
the press remarks.

He'll leave Saratoga Springs for Washington
to complete an apprenticeship
on the Board of Examiners-in-Chief,
judging whether patent applications
really do create something new.

Eunice tags along, thrust
from all she's known, the state
where her brain wrests free, soars
the cloudless cerulean of her imagination,

where she enlivens inventions from the depths
of consciousness. She'll settle in Washington,
where aftershocks of civil war & endless death
clap back, unsettled.

She'll settle in Washington.
In Washington, she'll settle.

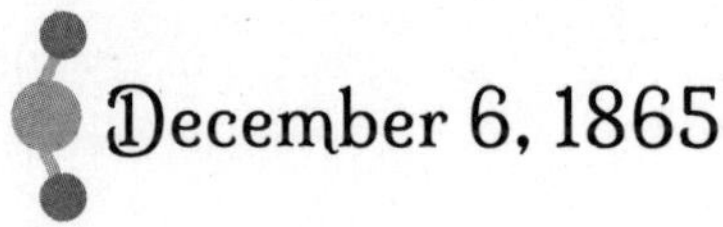

December 6, 1865

Denouement

The Georgia legislature
becomes the twenty-seventh state
to ratify the Thirteenth Amendment,
crossing the three-fourths threshold
to set four million Black Americans
                                        f r e e.

Children Now Grown

After two decades of work, Eunice
has achieved her greatest invention:
two grown daughters with minds for

        Fairness,
        Equality,
        Liberty
                for Women and for All.

She has lit their torches.
They must choose
whether to hoist the light.

Chances look favorable when
after moving to Washington,
Mary stands in the Senate chamber
balcony & flirts her way into a meeting

with a lawmaker on the floor below.
Senator John Brooks Henderson,
the rags-to-riches Missouri boy
who coauthored the Thirteenth Amendment.

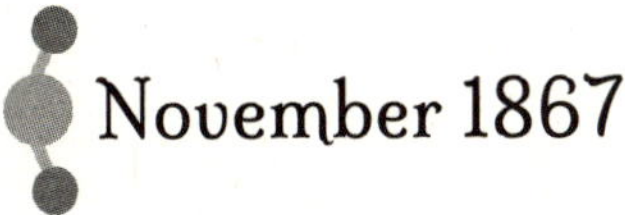

November 1867

A Wrap

A Fitchburg, Massachusetts,
factory has given Eunice
& her papermaking machine
a chance—in her name.

With this news report,
she can exhale
all the carbonic acid gas
her lungs have been holding.

The wrapping-paper factory
is saving $157 per day
in *raw material consumed,*
thanks to Eunice's invention.

No matter what,
Fitchburg will be etched
in the history
of Eunice.

1867

Fluid

Fixed air—
carbonic acid gas—
earns a new name,

carbon dioxide,
for its single carbon atom
combined with two oxygen atoms.

A new name
for the same substance formed
from burning organic matter—

formed because carbon,
the stuff of stars, needs oxygen
to breathe in the sky.

March 19, 1867

Epicenter

A noteworthy application
has come through the patent office:
No. 63,130, Burning Hydrocarbon

by one Henry Rutger Foote—
Elisha's younger brother
by fifteen years.

He's living in Oil City, Pennsylvania,
a fitting location for the inventor
of a new steamboat engine—
the very first to be powered

by flames, not of wood or coal

but of CRUDE
PETROLEUM
OIL.

By such invention I am enabled
to obtain practical results
of great importance and value,

Henry writes.

His apparatus of

tanks, coils, valves,
pumps, pistons, burners,

forged of coal-fired metal,
linking an already inextricable
web of *carbonaceous* materials—
fossil fuels burning

productive
powerful
heat.

Black Hole

Henry Foote acknowledges
recent experiments by the United States
Board of Commissioners showing
crude oil can be used as a fuel.
It is perfectly
safe
& economical—
more steam for less money
than coal.

Eunice's news about carbon dioxide,
about an untenably warm atmosphere,

goes continually unnoticed
& the family finds itself

at the epicenter
of the greenhouse,

just under the glass ceiling,
pressure building,

releasing CO_2 & discovering
the gas's blanketing properties

in tandem.

What Next?

It has been demonstrated that
oil can help manufacture pianos.
Now, it shall transform marine engines.

Eight tons of shovel-hungry coal,
replaced by
four small barrels of oil,
supplied only as fast
as it can be consumed
Henry Foote writes.

No more smoke and choking ash.
No more heaping piles of rock.
No more firemen or coal passers.

Science was arming another power
in the great laboratory of nature.

If a success, then a revolution must follow—
a revolution of no less magnitude
than that created by steam itself.

Man—one man—

was making history
for a proud and powerful nation,
and for the world.

Increase the intensity.

Spread the flame.

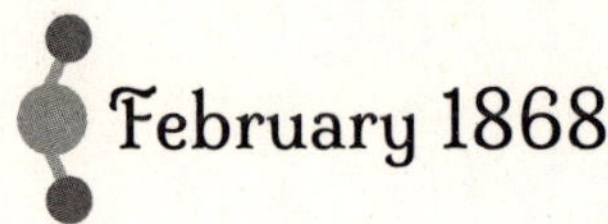

February 1868

The President Impeached

After months of controversy
among the men in power,
conflict embroils the country

again.

Hear ye! Hear ye!

In the stuffy, crowded Senate chamber,
the seventeenth US president,
Andrew Johnson, who filled
Abraham Lincoln's boots
at the pierce of a bullet, is hereby

IMPEACHED
for firing the secretary of war
without Senate approval,
a requirement Johnson
claimed was unconstitutional.

IMPEACHED.
The first commander in chief
to earn this dubious honor
(*wretched man,*
scoffs Rep. Thaddeus Stevens).

IMPEACHED.
Despite his loyalty to the Union,
despite betraying his home state
    of Tennessee when it seceded,
despite running for vice president
    as a Democrat—party of states' rights,
    party of slavery—alongside Lincoln,
    a Republican, to unite opposing sides.

IMPEACHED.
After House Republicans from Lincoln's party
jostled for years to reach this moment,
& now finally send charges to a Senate court.

Senate Trial

Present arguments.
Deliberate.
Decide

& President Johnson is hereby

ACQUITTED!

By a single vote!

One senator,
    John Brooks Henderson,
    friend of Lincoln,

listened to his fiancée,
changed his mind,
broke ranks with Republicans,
defended the president.

All with his wedding afoot.

June 20, 1868

Mary's Wedding

At last—something to celebrate!
The wedding of the century!

The dashing & accomplished
Senator John B. Henderson of Missouri
    (the enslaver who coauthored
    the Thirteenth Amendment
    to end enslavement & who
    rescued Johnson's presidency),

weds the fashionable young socialite
Miss Mary Foote, belle of Washington,
    daughter of Mr. Elisha Foote.

The press names a Mrs. Foote
as mother of the bride, a footenote,

defined not by her discoveries
but by her domesticity,

clad in exquisite lilac silk
that can scarcely compete

with her shimmering pride.

Adorned

The National Hotel ballroom hums
with fragrant white lilies,
draped lace
& distinguished guests
who fill the room with
celebri-fied carbon dioxide:

President Andrew Johnson
(lame duck redeemed),
General Ulysses Grant
(running to replace Johnson),
Supreme Court Chief Justice Samuel Chase
(praised by Democrats & scorned
by Republicans for presiding over
the impeachment trial)
& most of the Senate,
recessed for this event!

These men,
these gatekeepers,
can't peel their eyes off Mary

in her gown with full train,
in her crown of orange blossoms,
in her something-old fashion

& her something-new passion.

Gap

Few can ignore that twenty-five-year-old Mary
weds a man sixteen years her senior,
who entered business when Mary turned six.

Eunice nibbles on the day's delicious meal,
perhaps with a side of questions:

> Is this what it means
> for a woman to arrive?
> To leave her name behind?

Without the vote,
without science,
without her name—

> is it all
> enough?

Wedding II

While lawmakers swill decadence
& rejoice in nuptial revelry,
the nation sways,

days from ratifying the Fourteenth
Amendment to the Constitution, which
many at this event support, which gives
citizenship & civil rights to Black people.

They mutter of a Fifteenth Amendment
proposed for next session's docket,
which would grant the vote to Black
                                        men.

More than the who's who of the guest list
& the what's what on the gift table
(silver ladles,
        gold bracelets,
                gleaming dessert spoons,
                        suffrage for the select),

more than what Eunice & women before
have been unable to accomplish,

more than the transparent ceiling
she's been unable to kick through,

the biggest gift Eunice can offer her daughters
is the fire fueled by fossilized expectations,

the will to watch them burn
& the prescience to pass the torch.

Carbon

Let it sear the ceiling.

Let cracks pervade generations.

Let the whole shell cave.

Landscape

Eunice emulates the masters
in her panoramic oils

as witness to Earth's gradient
of blues, greens & browns,

a muddy mess in the wrong hands,
a vibrant faith in the righteous ones.

Opportunity

One of the wedding guests
extends the offer of a lifetime   to Elisha.
President Johnson himself
sees potential in the father of the bride.

Elisha has been paying his dues
in the US Patent Office the last three years.
Applications have doubled since
the end of the war. Elisha has filed

his own patents, ushered in untold others
& as an attorney, defended inventions
before the one & only Supreme Court.
A month after the Foote-Henderson wedding,

in the wake of a corruption scandal
involving the head of the patent office,
Johnson has a decision to make.
He nominates Judge Elisha Foote

to the helm as commissioner.
Imagine—
commissioner of patents!
For the whole hobbled country!

Patent Office

Color Eunice pleased
with how far she's come:

Discoverer of the warming
properties of carbon dioxide.

Woman scientist onstage among men.

Holder of two patents in her name—
squeak-free rubber insoles
& a papermaking machine—
& a secret third, her parlor stove.

Defender of women's rights.

Wife of the patent commissioner.

Mother of two strong women.

She guides a reporter
through the patent office—

Elizabeth Cady Stanton herself!

Libby writes for and edits a new
weekly women's-rights newspaper

published by Susan B. Anthony—
The Revolution.

Three decades of construction
have just completed the patent office—

gas lighting & running water,
skylights & marble floors.

The tour highlights
the *museum of curiosities*—

rows of glass cases cradling
thousands of model inventions,

including Eunice's own.

Engrossed

Eunice breathes in the history—
not just of the nation's creativity

but of the nation. These colonnades
housed Union soldiers resting their heads

before their next battle. Nursed
mortally wounded troops

languishing between display cases.
And after the hospital closed,

celebrated the Lincolns, waltzing
at their second inaugural ball

past the boat-buoy patent model
whittled by Honest Abe's hands.

Libby interrupts:

Why

aren't there more
women inventors?

Honest

Eunice knows she can speak
candidly. Libby is a friendly audience.

Ever since the Stantons left Seneca Falls
for Brooklyn in the early sixties,

Eunice has seen Libby's name everywhere,
still linked to the women's cause.

The Revolution's articles even appear
over the wires, despite its fiery slogan:

PRINCIPLE, NOT POLICY:
JUSTICE, NOT FAVORS.—

MEN, THEIR RIGHTS
AND NOTHING MORE:

WOMEN, THEIR RIGHTS
AND NOTHING LESS.

Eunice lifts kindling
to Libby's match.

Illumination

No doubt, Eunice says,

that half the patents there
were the inventions
of women.

She blows gently
at the spark.

Men
had the money,
loved notoriety,
took out the women's names,

Eunice says,

& thus snuffs
the gaslighting
& sets the torch

ablaze.

The Newton family homestead still stands near East Bloomfield, New York.

Emma Willard, founder of the Troy Female Seminary, first in the US to offer higher education—including science—to girls.

The Troy Female Seminary science lab, where Eunice learned the scientific method.

Eunice N. Foote

While there is no confirmed image of Eunice, documents show her signature.

Judge Elisha Foote, Eunice's husband.

Mary Foote Henderson, Eunice's first child.

Senator John Brooks Henderson, Mary's husband.

Augusta Foote Arnold, Eunice's second child.

No one knows for sure what Eunice looked like . . . but could this be her? Tompkins H. Matteson depicts the 1856 dedication of an observatory in Albany, New York, which coincided with a conference Eunice and Elisha attended. (See p. 268 for more information.)

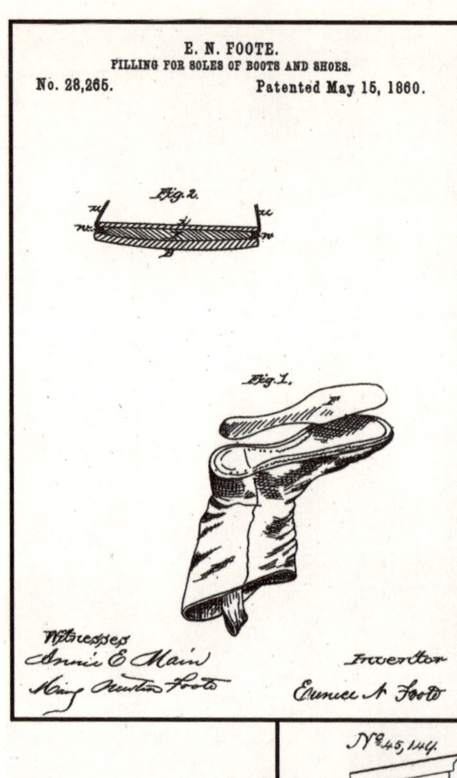

Eunice's patented rubber insoles prevented shoes from squeaking.

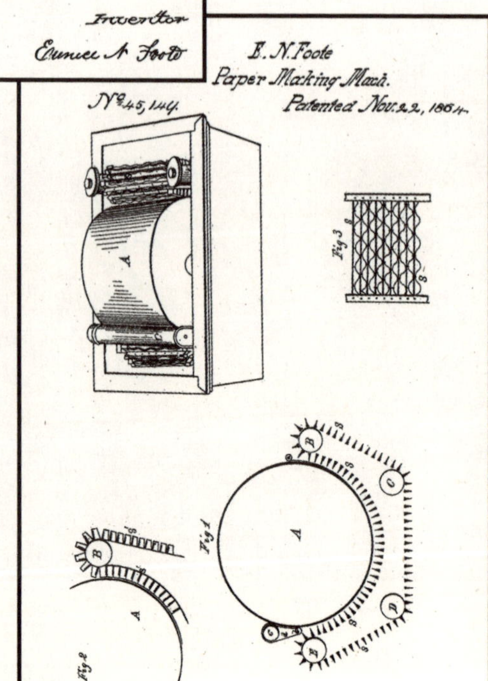

Eunice invented a machine that strengthened wrapping paper.

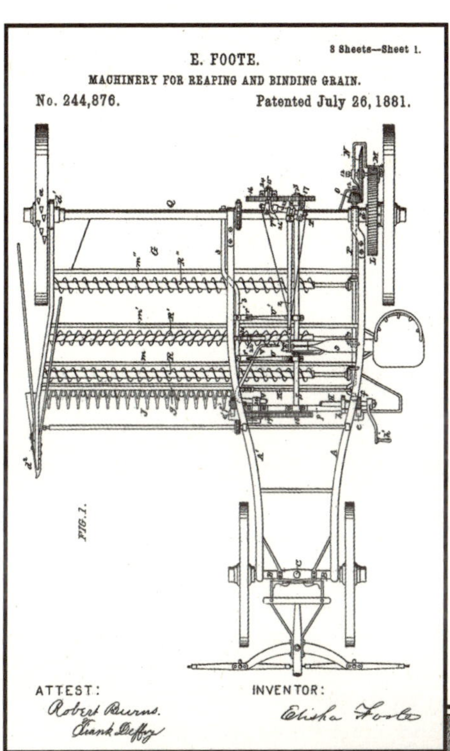

Elisha created a self-binding reaper, which simplified harvest for grain farmers.

A "museum of curiosities" at the United States Patent Office showcased thousands of patent models (shown here in 1856).

382 *On the Heat in the Sun's Rays.*

ART. XXXI.—*Circumstances affecting the Heat of the Sun's Rays;* by EUNICE FOOTE.

(Read before the American Association, August 23d, 1856.)

MY investigations have had for their object to determine the different circumstances that affect the thermal action of the rays of light that proceed from the sun.

Several results have been obtained.

First. The action increases with the density of the air, and is diminished as it becomes more rarified.

The experiments were made with an air-pump and two cylindrical receivers of the same size, about four inches in diameter and thirty in length. In each were placed two thermometers, and the air was exhausted from one and condensed in the other. After both had acquired the same temperature they were placed in the sun, side by side, and while the action of the sun's rays rose to 110° in the condensed tube, it attained only 88° in the other. I had no means at hand of measuring the degree of condensation or rarefaction.

The observations taken once in two or three minutes, were as follows:

Exhausted Tube		Condensed Tube.	
In shade.	In sun.	In shade.	In sun.
75	80	75	80
76	82	78	95
80	82	80	100
83	86	82	105
84	88	85	110

This circumstance must affect the power of the sun's rays in different places, and contribute to produce their feeble action on the summits of lofty mountains.

Secondly. The action of the sun's rays was found to be greater in moist than in dry air.

In one of the receivers the air was saturated with moisture—in the other it was dried by the use of chlorid of calcium.

Both were placed in the sun as before and the result was as follows:

Dry Air.		Damp Air.	
In shade.	In sun.	In shade.	In sun.
75	75	75	75
78	88	78	90
82	102	82	106
82	104	82	110
82	105	82	114
88	108	92	120

Eunice's consequential 1856 experiment identified the warming properties of carbon dioxide, which she knew as "carbonic acid gas."

The high temperature of moist air has frequently been observed. Who has not experienced the burning heat of the sun that precedes a summer's shower? The isothermal lines will, I think, be found to be much affected by the different degrees of moisture in different places.

Thirdly. The highest effect of the sun's rays I have found to be in carbonic acid gas.

One of the receivers was filled with it, the other with common air, and the result was as follows:

In Common Air.		In Carbonic Acid Gas.	
In shade.	In sun.	In shade.	In sun.
80	90	80	90
81	94	84	100
80	99	84	110
81	100	85	120

The receiver containing the gas became itself much heated—very sensibly more so than the other—and on being removed, it was many times as long in cooling.

An atmosphere of that gas would give to our earth a high temperature; and if as some suppose, at one period of its history the air had mixed with it a larger proportion than at present, an increased temperature from its own action as well as from increased weight must have necessarily resulted.

On comparing the sun's heat in different gases, I found it to be in hydrogen gas, 104°; in common air, 106°; in oxygen gas, 108°; and in carbonic acid gas, 125°.

Art. XXXII.—*Review of a portion of the Geological Map of the United States and British Provinces by Jules Marcou;** by William P. Blake.

Geological maps of the United States published in Europe and widely circulated among European geologists, are necessarily regarded by us with no small degree of attention and curiosity. This is more especially true, when such maps embrace regions of which the geography has only recently been made known and the geology has never before been laid down on a map with any approach to accuracy.

The recent geological map and profile by M. J. Marcou, which has appeared in the Annales des Mines and in the Bulletin of

* Carte Géologique des Etats-Unis et des Provinces Anglaises de l'Amérique du Nord par Jules Marcou. *Annales des Mines,* 5e Série, T. vii, p. 329. Published also with the following:

Résumé explicatif d'une carte géologique des Etats-Unis et des provinces anglaises de l'Amérique du Nord, avec un profil géologique allant de la vallée du Mississippi aux côtes du Pacifique, et une planche de fossiles, par M. Jules Marcou *Bulletin de la Société Géologique de France.* Mai, 1855, p. 813.

Frederick Douglass, former enslaved man and abolitionist.

Elizabeth Cady Stanton, women's rights advocate and leader.

Joseph Henry, professor and first head of the Smithsonian Institution.

William H. Seward, US senator and then secretary of state.

Part Three

These women were then unpopular. Although their lives were so blameless as to defy calamity, the weapons of ridicule were substituted, and used against them with relentless power.

—Mary Foote Henderson,
published author,
February 16, 1890

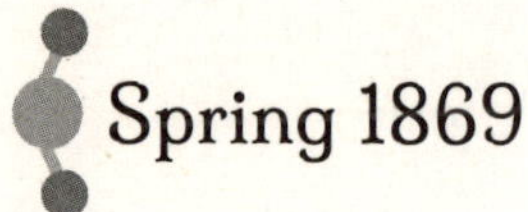

Spring 1869

The End

When General Ulysses Grant
assumes the presidency,
he opts to replace Elisha
as commissioner of patents,

despite a strong letter
of recommendation from
Elisha's esteemed friend
Professor Joseph Henry.

While in office,
Elisha oversaw thousands of new patents
& encouraged the use of new inventions—
photolithography, for one,

using stone & chemicals
to create backup copies
of every patent drawing.

But first,
before he leaves,
Elisha must right
a wrong.

For Years

Women have kept
the patent office machine
grinding along

as copyists working from home,
churning out pages, paid a dime
for every hundred words scrawled.

Just before Elisha leaves office,
Congress allows fifty-three women
to slide into desks in the building.

But Elisha soon learns
they earn only $700 each year,
while male copyists make $900.

Eunice has shown Elisha
the value of women's work,
so he cuts several male clerks

to clear space for equal pay.
To clear space
                    for justice.

March 6, 1869

Frankly

Another year, another wedding.

Augusta has chosen for her venue
the First Presbyterian Church.
Augusta might have liked that it was
one of the only Washington pulpits
to welcome Fredrick Douglass after war's end.

Augusta's Washington nuptials
to the wealthy tea & coffee merchant
Francis Benjamin Arnold—Frank—
attract less attention in the press
than Mary's wedding & fewer
notable dignitaries.

He treats Augusta well
& fluffs the cloud she floats on.
The couple will settle in New York City
for lavish apartment living.

She'll busy herself filling
every bare space on every wall,
coating the parlor in crimson
& planning a family.

With that, Eunice's nest

officially empties.

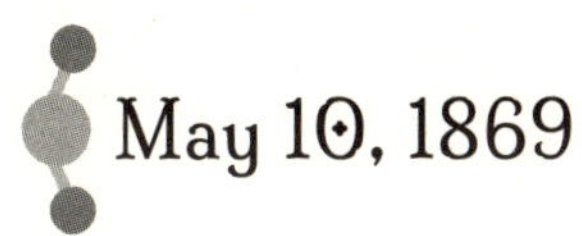

May 10, 1869

Transcontinental Railroad

2:27 p.m.

Telegraph offices across America
fall as silent as spring snow.

2,400 miles west of Washington,
at Promontory Point, Utah Territory,
a signal dispatches:

Almost ready.
Hats off; prayer
is being offered.

2:40 p.m.

The telegraph bell sounds.

The spike is about to be presented.
The signal will be three dots
for the commencement of the blows.

2:47 p.m.

Hammer raps golden spike.
Magnetic hammers tap bells
across the land:

dot-dot-dot
Done.

Confluence

As twin rails link Atlantic to Pacific,
as colonizers & carbon ooze like a wound,
as cities parade & dance & light fireworks,

a new era begins.

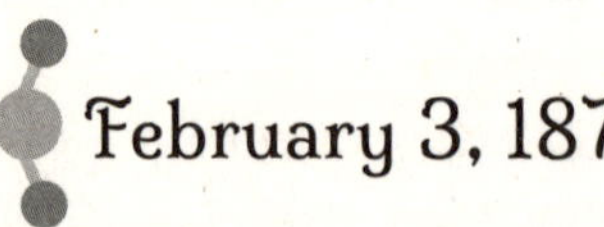

February 3, 1870

Black Men Win the Vote

The Fifteenth Amendment
gets its signature
from President Grant,

& thus
Black men win
the right to vote.

It's vile not to cheer
the granting of rights
to marginalized people.

Perhaps Eunice thinks
of those who fought
so hard for this moment

& of those who have fought
for twenty-two years now
for the moments to come.

When
will women
get their moment?

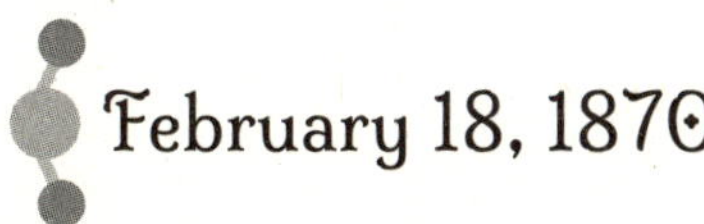

February 18, 1870

Eunice's Beginning

A new era for Eunice
begins with her transition

from star of the Foote solar system
        to satellite,

from Mother to Gram.

The journey carries turbulence;
Mary's water stays inside far too long.

I wonder you did not die,
                    Eunice reflects.

But like a miracle from the sky,
precious John Henderson II—
        Johnny—

makes his grand entrance
in his own time,

a world away from Washington
in a small Eastern Missouri river town.

As the season shifts, Eunice rejoices,
knowing who will now carry the torch,

alighting her dreams
for her family & the world.

March 1870

Alone

Out here in the middle—
in tiny Louisiana, Missouri—
so far removed from everything,

Mary spirals,
sleepless
as baby wails
& flails,
as Mary fails
to squeeze
precious drops
of liquid gold
to keep little Johnny
alive & thriving.

Halfway across the continent,
Eunice anguishes, confesses
she cannot come to Missouri—
cannot ease Mary's difficulties
with the newborn—yet.

John has gone—resumed his travel—
but between business & politics
he finds time to send help
for Mary—a wet nurse.

The young woman's miraculous milk
will stream sustenance for the baby
& make up for what Mary lacks.

Generations Bloom

The magic hour before twilight
casts long shadows in golden sun.

With family flung in all directions,
futures shimmer like unframed paintings.

Mary, John & baby Johnny
have left for Saint Louis.

Newlyweds Augusta & Frank
indulge in NYC theater & dinners

as Gus implores Mother to hurry
before baby takes the stage.

Elisha is buttoning up work
in Troy & Washington,

while Eunice has been
biding time in East Bloomfield.

So Much Has Changed

Since the first days on this farm,

seedlings of equality have set fruit
in the ripening suffrage movement.

Mandy—Aunty—has grown gray,
managing the land against the odds.

Eunice once rose above these trees
like a hot-air balloon,

steering an unlikely career
as scientist & inventor,

navigating currents
as wife & mother.

Twilight cools the air.

Eunice descends back to now
as Gram, puffed & proud as ever.

Eunice Frets

How much advice
should she give
her grown daughters?

Sweet baby Johnny is not yet creeping,
not pulling up, not staying as quiet
as Eunice believes a nine-month-old should.

The rides in his little wagon—
          do they hurt him?

The attention Mary pays him—
          does it spoil him?

She scrawls a letter to Saint Louis:

I wish you would get his picture taken
& send it to me, weigh him,
& measure his length
& send it to me.

The &s could fill a train to Missouri,
but Eunice prunes her angst
& seals it with love:

Aff—

Affectionately—

Mother

Cycle

Butterflies dance
          in late-morning sun.
                    Earth's heartbeat sounds
                              on the cricket's wing.
                    Eunice packs her trunk,
          bids the farm goodbye
& takes leave eastward
          to usher grandchild No. 2
                    into the changing world.

Nesting

They call it a sickness,
this condition of delivering
new life from one's womb.

Just one of many sicknesses
women must endure. Gus
has been readying her nest,

setting out lace that winged
from Mary in Saint Louis
to New York, feeling every flutter
of the fetus inside her.

Eunice bobs about New York City,
collecting decorations for
Mary's new home in Saint Louis

& ingredients for the catnip tea & barley water
she believes baby will need in the hours, days,
before Augusta's milk begins to flow.

After all this time,
Eunice's fledglings flown & grown
with nests of their own,

she still basks in
her most treasured role:
Mother.

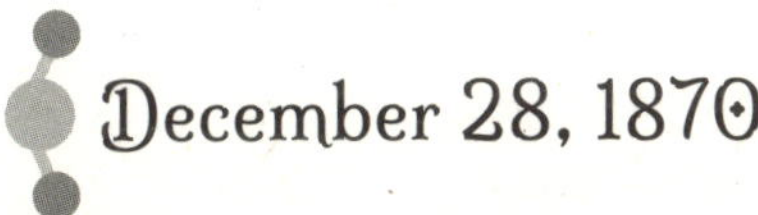

December 28, 1870

At Last

There is news to share:
Gus has gifted the family
with another grandson.

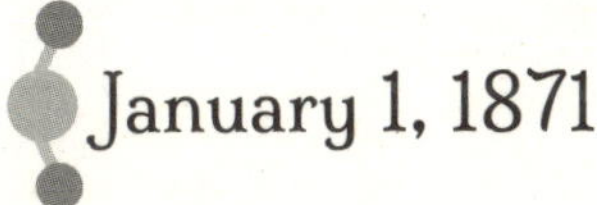

January 1, 1871

Letter

Happy New Year to you
My darling child,

Eunice writes to Mary.

Benjamin is four days old
& is the best little fellow
don't cry or make any trouble

I expect he is afraid
of his Gram's displeasure
if he makes a fus

Guss is getting on well
is inclined to cry
if she gets tired—

but we give her something
to stimulate & all her bad feelings
pass off & we have sunshine again

she was only sick 12 hours
took chloroform & was not concious
of much suffering—the discovery

of that remedy is a great blessing—
she has a most excellent nurse
& good Dr & will come out

all straight in a few days

The day
the month,
the year,
the life,
is charming.

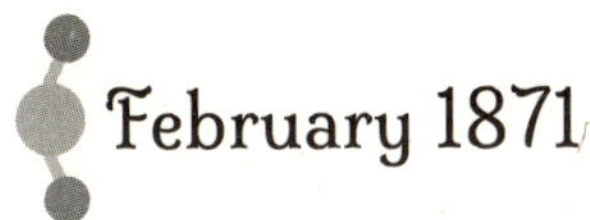

February 1871

Babies

Inhaling that new-baby smell,
        fluttering kisses on fine hair,
                tickling baby toes

must end. Eunice swaps time
with one grandchild for another,
leaving Gussy's baby, Benjamin,
        the finest boy there ever was,

to join Elisha in Washington.
They'll crisscross the Confederacy
chasing the other half of her heart.

Mary's little Johnny has grown—
she's finally mailed a photograph—
and Eunice can't pack her trunk
fast enough. Questions swirl:

        How does he babble?

                What are his interests?

                        Does he remember his Gram?

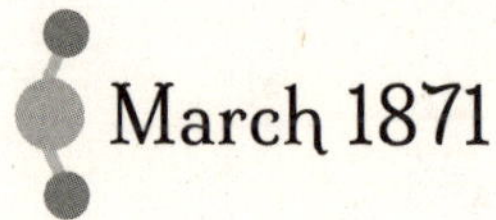

March 1871

Steamboat

Eunice & Elisha's excursion
begins with a sojourn in Charleston,
that beauty of a rebounding Southern city,

& continues through Dixieland—
emptied of Native peoples forced
to trek out on foot though the Footes

have the privilege of gliding.
In New Orleans, gateway
to the waters of the middle,

they board a floating birthday cake
called the *Olive Branch*.
The sidewheel steamer slices

north along the Mississippi—
first through Louisiana's sugar bowl.

Eunice devours the landscape
with a scientist's curiosity.

Sickly Sweetness

Black arms glint as machetes

chop,

chop,

chop

ten-foot stalks of sugarcane,
hungering toward the same sun
that starves the life from workers
keeping pace with a ravenous mill.

The juice of these stalks
extracted by machine,
rollers powered by steam,
guarded by swords
for slicing off any limbs
inconveniently pulled in.

Juice boils down
to coveted sugar crystals.
The machine of slavery
has evolved to sharecropping.

The machine of antebellum
human- & horsepower,
sweetened with blood,
evolved to freedmen
& advancements of
belching, choking coal.

Rebuilding
after war's decimation.

Reconstruction,
they call it.

Reimagining
the same old
systems of power.

Cotton

As the *Olive Branch* steams northward,
the riverside farmscape shifts from sugar

to cotton.

Cotton picked by
bleeding hands
& hunched backs,
working harder
than any backs
should work.

Cotton ginned, spun into

layers
          upon layers
                    of cloth—
                    the kind
          that wraps
Eunice's body.

Layers

Delicate stockings,
chemise undergarment,
underdrawers,
steel-boned corset

-cinch & fasten-

bustle,
petticoat (a cloud of ruffles),
corset cover

-button-

a vibrant gown—
first the skirt

-button, button-

then the bodice

-button-

finished with lace & lace & lace,
accentuating the feminine bust,
reminding admirers that

what is cultivated

must be

controlled.

Incendiary

Eunice thrills at stolen glimpses
of dinosaur's cousin alligator—
lurking in the Mississippi's waters,
peeking, blessing the air Eunice inhales.

What a wonder that steam navigation
has become possible in Eunice's lifetime.
The industry waxed—
opened the West to Europeans—
& now waned, with rails proving safer.

And Lo!

FIRE!

S m o k e pours

from one of the rooms,

c l o u d s the hull,

descends this packet steamer

into

CHAOS.

But Soon

After a mere coughing spell,
the flames are extinguished.

Eunice writes to Mary:

I will
          be with you
soon.

July 1871

Safe in Saint Louis

Humidity hangs above the confluence
of the Mississippi & Ol' Muddy.

Cicadas sing their chorus, unperturbed
by the bombshell in today's paper—

the Missouri River has swallowed
a packet steamer northeast of Kansas City.

The elegant *Olive Branch* proved too weak
to appease the stump that snagged her.

Wait! The *Olive Branch*—that's the boat
Eunice and Elisha just disembarked from!

No lives perished besides a few rats,
but four feet of murk spoiled the perishables—

flour in the hold, bran & beer on the deck.

The ship, she be *Gutted,*
said man.

Above the Fold

As Eunice inhales the sultry air,
perhaps she remembers
the scent of soot clinging

to cabin drapes after the boat fire.
Perhaps she praises the creator
for delivering her & Elisha

to Saint Louis in one piece.
Perhaps disaster lends trepidation
to the Footes' eventual trek home.

Fortunately, there's no rush to leave,
for Mary has written a headline
more momentous than any sunken ship:

Eunice Will Soon Be
Gram Times Three!

August 1871

Babymoon

With Eunice on hand to fluff Mary's nest
in Saint Louis, Mary sheds the cloak of heat
& mosquitoes for an eastern sea breeze.

She coasts into her third trimester
for baby No. 2 at her new cottage
in Long Branch, New Jersey,
the elite beachside retreat

of His Excellency, President Grant.
While Mary revels in Grant sightings,
Eunice tinkers away on the third story
of the Hendersons' new home,
installing carpets, beds, bedsteads—

& there's no chance she will allow
Mr. Henderson to foot the bill.
He's got enough on his mind

as he grows his legal practice
& dreams up new ways
to dote on his growing family.

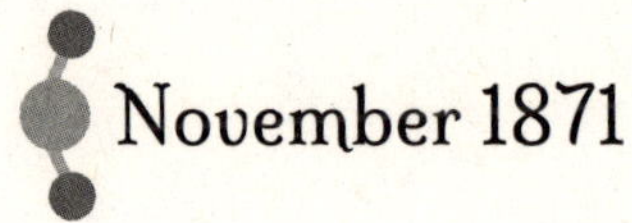

Maud

Something about
the second time
sends waves of calm
over a family.

The second time,
Mary knows
what to expect.

The second time,
Eunice knows
the first baby survived.

The second time,
a life-giving, breathing
GIRL emerges,

& her name almost
chooses itself:

Maud,
meaning
might in battle.

John Writes to Mary

Once again, John has left Mary
alone in Missouri with a newborn.
As she navigates little Johnny's

possible exposure to scarlet fever—
a scourge worse than cholera or smallpox—
her husband works in Washington,

invoking his new role as candidate
for Missouri governor. John meets
Elisha in the city & enlists him
to visit Saint Louis & play caregiver.

John writes,

I think he will consent to go out
and stay with your mother
and the children for a few weeks.

But how will the mother take it?
Does she consent? You seem
to assume that. Poor mother.

She markets and looks after the children
and the house. Instead of a philosopher,
a scientist, an inventor,
a builder of water machines,

she has become a mere drudge,
a servant of servants,
worse than all, a housekeeper.

Will she stand it?

She Will

As a star stands
the dust of nebula,

as a mother bird stands
her young's gasping beaks,

Eunice will more than
stand her expanded role
with her grandchildren.

She will embrace it.

March 1872

Back to East Bloomfield

Elisha continues to chase
invention after invention.

This time, Patent No. 124,944
makes an *Improvement in Driers,*

whether for fruit, vegetables,
grain, wool, or any other purpose.

But Eunice has a different
purpose on her mind.

Legacy

What will Eunice impart
to the next generation?

Augusta is with child again,
& Gram is needed
everywhere.

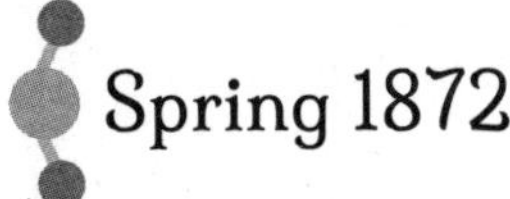

It's a Boy!

Augusta has safely delivered
the world's newest human:

Francis Arnold, Junior,
named after his father.

A sweet boy—
what joy!

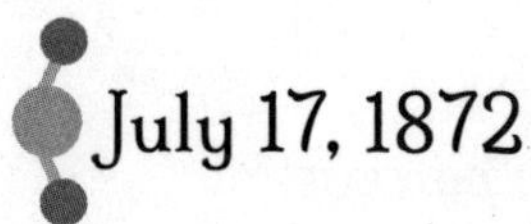

July 17, 1872

Changes

Eunice remains in Saint Louis with Mary,
but she fears overstaying her welcome.

She will leave when John returns.
But there is more to consider
when Mary shares John's appeal:

My Dear Wifey

She is not only welcome, but she is now
a necessary member of our little house hold.

Indeed she must stay with us and you must
communicate any wishes to her at once.

Our days on earth are short, and we
should so use them as to give ourselves
most happiness.

Hubby

France

Augusta wastes little time healing
before she & Frank follow through
with plans to flit about France.

They tote along the wee ones—
baby Francis with his adora-dou-ble chin
and bracelet-like wrist rolls.

The best baby I have ever seen,
she writes,

& Benny, acting like *quite a Frenchman*
in his apron, digging in the sand,
understanding French chatter at

the loveliest seashore place
I have ever seen,

crescent shaped, smooth & safe
for bathing at low tide.

Moonlight on shimmering glass
gives way at dawn to colored sails, fanning
like paintings as vessels stalk sardines.

They've been to Paris—
a most fascinating place
and hard to leave,

& now the family will remain seaside
another few weeks, visiting *châteaux*,
watching peasants in crimped caps & ruffles,

perhaps gambling, tasting decadent fruits
(strawberries, apricots, pears,
plums, grapes, almonds, figs . . .)

& hoping Mother, Father & Mary
will hurry up & write back—
or better yet,
set sail.

Labor

The people here work very hard
women as well as men,

Augusta writes.

The former do the same & as much
as the latter—drive carts—
pitch hay—carry burdens &c.

It makes me feel badly to see
human beings work so hard.

The men farmers themselves in carts
and you see them laboring so to draw
the load they nearly are bent double.

Mary has not written me
and you have all been remiss
except Mother. Mend your ways

& write often
to yours Aff'y—

Gus—

Oh Là Là

Augusta's dream unfolds in letter after letter,
only to be met with silence from her family.

She describes
  watching beach races from her balcony,
    plotting a float down the Rhine,
      fetching cheap French fashions.

I hear you are suffering very much
with hot weather,
she closes,

wondering
  if she'll hear anything
    in return.

In Fact

The family is suffering.
But not from hot weather.

August 1872

Letters to Long Branch

Monday, August 12

I am uneasy and will be uneasy
about the children till I hear from you,

John writes to Mary.

If either child becomes very sick,
you must telegraph me,
and I will endeavor to be with you at once.

Saturday, August 17

My Dear Wifey

What is the matter? Why don't you write?
I am so uneasy about the children.

Affectionately yours
Hubby
Kiss our babies for Papa

Monday, August 19

I am so uneasy about the baby,
but I trust you will spare no pains
and no expense to save her.

Your mother left Bloomfield
on Wednesday and possibly arrived
Thursday evening at the Branch.

I presume your father is with you,
as he was at Yonkers when I telegraphed him.

I have an appointment to speak on Thursday,
but I think I will not go there, but instead

leave for Long Branch on Thursday or Friday
reaching there about Saturday—

Carbon Cycle

All life is made of stardust.
Carbon lives & carbon dies, turns
to dust that sinks, buries deep, where
it must stay—until seed roots, until
roots stretch, until green emerges,
until animals consume, until

the cycle repeats.

Broken

It
should
not

be
natural

for
a
baby

to
die.

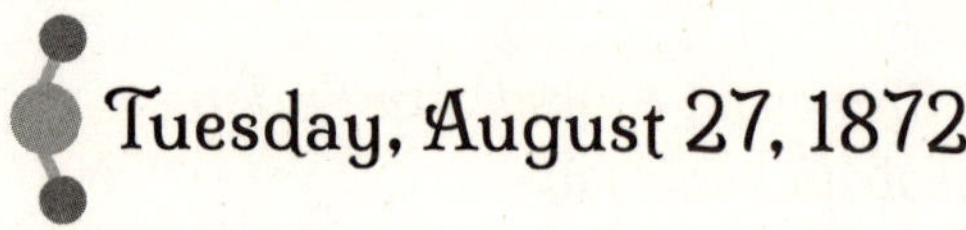

Maudlin

Nor should a baby girl's brain

s w e l l

from disease.

Yet that is how
Mary's daughter,
Eunice's grand

Maud,

dies at nine months:

congestion o f t h e b r a i n.

From Maud (Part XVIII)

There is none like her, none.

***

My own heart's heart,
my ownest own, farewell . . .

—Alfred, Lord Tennyson, 1855

It Happens So Fast

One day
the family is focused
on big John's campaign
for Missouri governor, breathing
the fresh New Jersey air;

the next day
Maud is gone;

the third day
she's a wisp of memory,
six feet
    under
    Brooklyn.

It So Happens

Elisha has been planning
for his own eternity.

Just two months earlier—
two months before his sixty-third birthday—

he tracked down three hundred feet of soil
in the prestigious Green-Wood Cemetery.

Stone by stone, the mausoleum
stacked up, its forehead etched with

FOOTE.

But this resting place for legacy
now cradles only grief,

for the first feet in this tomb
belong to a Henderson

& take up
very
little
room.

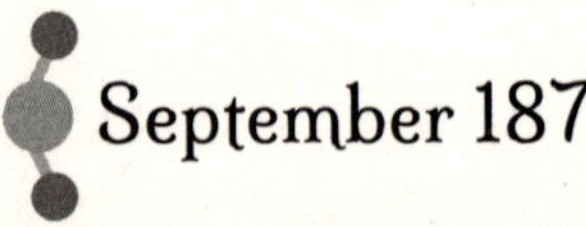

Back to the Farm

Eunice leans in as Gram,
whisks Johnny away
from the trauma,
sadness,
turmoil
roiling his Missouri home.

At two and a half, Johnny's a sponge
who needs space
& attention
& love
to grow.

So while his mother mourns
& his father finishes the final leg
of his gubernatorial campaign,

Eunice slows life to show Johnny

how to feel connection to the earth,

how to feel connection,

how to feel.

So Digging Potatoes, It Is

Discovering what lies beneath—
a world of life made of carbon.

Gardening looks like play
to the unspoiled eyes of a toddler,

& the boy grows so soiled he could
grow potatoes between his toes.

Which is exactly what
both of them need.

Harvest II

Reaper rattles behind a pair of horses,
slicing, flattening amber waves.

Elisha observes Mandy's crew,
summoning their strength
  to rake the wheat stalks,
    bend & wrestle them
      into shocks
        by hand.

Shouldn't reaping be easier?   he thinks,
echoed by the words from Augusta's letters:

It makes me feel badly
to see human beings work
so hard.

The cogs of inquiry creak to life,
& Elisha tinkers—
twisting wires,
rattling iron,
trying this & that
to modify the machine—
to reap efficiency
from farmers' toiling hands.

Healing & Growing

Johnny is growing so much,
                    Eunice writes,
that Mary must send new cool-season duds.

Poor little fellow must look like a pauper,
                    Mary responds
                    with amusements & apologies,

(she should have *brought in a sewing girl*
                    instead of

the French girl who cannot do a thing
about the little clothes when I am with her)

I will certainly get his things
off this week—stockings, hat & all.

Love to Auntie
Affec'ly
Mary

And by the Way

                              Mary pivots nonchalantly,
                              as if grief
                              has loosened its grip:

I received a letter from Gen. Grant
written by himself, which I will copy for you.

President Grant!

Eunice tries to imagine
hobnobbing with the likes
of the hero president who
won the war for the North.

Perhaps it was his wife,
First Lady Julia Grant,
who first read Mary's letter
& shared it with her husband.

As far as hubby is concerned,
Mary writes,

he would prefer a first class
foreign appointment to the Senate anyway.
And if the State goes as all the Democrats bet,
the legislature will not be so very Republican.

However there is a good chance
for the Senate but if it should fail, I know
hubby would consider it all very fortunate
if he could get the appointment, and any one
knows he could get that easily enough.

Perhaps Eunice wonders:
Have they given up on the governorship?

Mary Will Not Give Up

Eunice knows this.

There is the greatest enthusiasm
here now for hubby, Mary writes,

and you don't know what an amount
of attention I get every where
I go—in stores or anywhere.

All of the cards I get
          for millinery openings &c.
                    would make you laugh.

Eunice is surely tickled
at how far her own torchlight
could spread.

Meanwhile, Beyond the Sea

Augusta's words ring

with grisly prescience—

Paris is a most fascinating place

                              and hard

                    to leave.

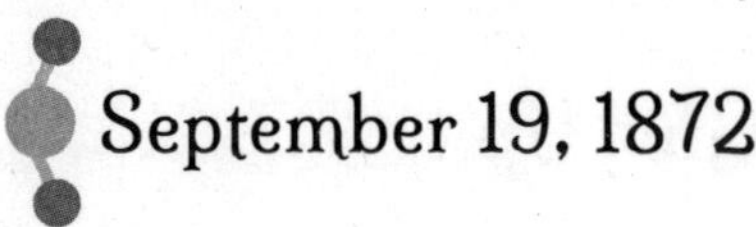

September 19, 1872

Francis Junior

Augusta's second child,
five months young,

dies suddenly,
less than a month
after his baby cousin.

Mort en Paris, oui.

City of fanned cobblestone,
gilded gargoyles
& angelic frescoes.

His chubby arms,
his cherubic face,
his perfect disposition

lie in perfect stillness—

Francis Arnold, Junior,

frozen in France

forever.

Empty

It
should
not
be
natural
to whisk
two
children
abroad
but
sail
home
with
only
one.

Two

Two weeks to journey
home on the *Scotia*.

Two daughters
in two states.

Two
empty cradles.

Two woeful holes
in Eunice's heart.

Hearts Beat On

Distract oneself
with purposeful work,
& holes of the flesh

can heal.

Scars may fade,
but rarely
do they disappear.

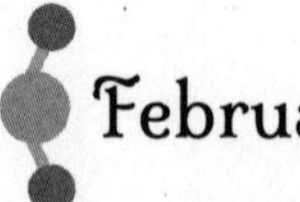

February 18, 1873

Patently

Elisha has retired,
but he'll never tire
of chasing dreams.

During his latest stint
on the farm, he harvested
another idea:

automated grain-bag ties
that simplify harvest
like an extra set of hands.

To deploy, affix a tear-shaped wire
  to a thin rope of manila, jute, or hemp.
    Tighten the rope around a stalk bundle.
  Thread it through the curved metal loop,
et voilà! The wheat shock is secure.

Be it known
that these binders
of Patent No. 135,899
comprise but one component
of a bigger machine to come:

Elisha's self-binding reaper.

1873

Unbound

Speaking of patents,
the US Patent Office has hired
its first female patent examiner,
a woman named Sarah J. Noyes.

May her eagle eyes,
or talons, perhaps,
reap inventions
from more women.

Spring 1873

Reunited

Three years ago
Augusta wrote,

I think it is a dreadful
shame you have never
been to visit me in NY

& mailed Mary a painting of the Madonna
& advised her to cloak her parlor in red
& begged her to visit—

soon.

And as Maud grew in the womb,
Mary ventured to New York to see Gussy.
But two more years have passed.

How? How has the sun set
on hundreds more days
& on Mary's child & Augusta's?

As Mary wends toward the city
to revive her sisterhood—

a possibility now that John
has lost his campaign for governor—

she feels her swelling belly
& sees the blind promise

of hope.

Harmony

How fitting that
Mary's joyful song
of a fetal sequel

coincides with
Augusta's reveal
of her own blessing—

a cousin due
come autumn!

April 15, 1873

Witness

Mary has journeyed to bear witness
to Elisha's latest patent application.

He's staying in Yonkers now,
just a hop and a skip away from Augusta.

Mary signs her name—

M. F. Henderson—

with all the confidence
her mother has instilled,
to Patent No. 137,905,
an improvement
to gas burners
that eliminates flicker,
that conserves the fuel
descended from dinosaurs.

It becomes heated, Elisha writes,
and communicates its heat
to the rest of the burner
and to the gas that flows along it,
and a whiter light
and more perfect combustion
are thereby produced.

Communicating heat
produces brighter light
& more perfect
combustion.

Just like the torch
Eunice intends to pass.

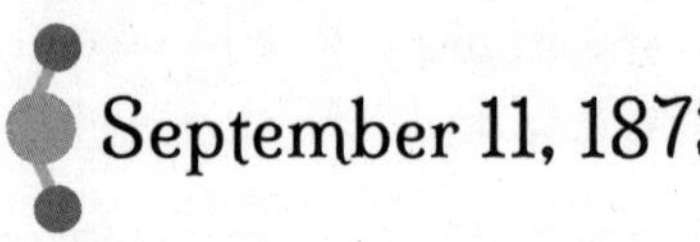

September 11, 1873

Sweet Joy, It's a Boy

Gussy safely breaks
through another sickness,

injecting life & joy where
once there was none.

Benny reclaims his title
as big brother, barely recalling
when a baby's cries last filled
the Arnold house almost a year ago.
Faded wisps of memory
                                                            replaced
by new words & skills & everything
an almost-three-year-old must learn.

This new child—

                    Henry Newton Arnold—

will not supplant
the life Paris claimed.

But he will grow
his family's hearts
just the same.

Mary's Turn

Halfway across the continent,
a world away from New York,

another boy screams to life,
born into privilege & wealth,

but no promise of health.
Sweating, exhausted, hopeful,

Mary anoints him
Francis Arnold Henderson

in homage to his departed cousin.

She'll call him

Arny.

Bouncing

The hands of the clock spin the year away

while Eunice
divides time
between
stretching,
grasping,
traveling

to keep the pieces of her heart together.

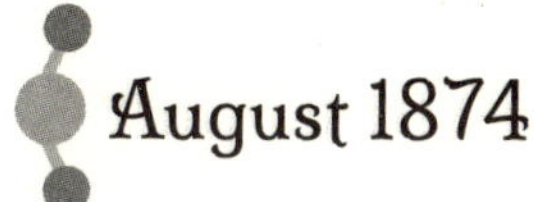

August 1874

Mary

Mary's letters reach Eunice
in New York, where the anticipation
of Augusta's next blessing, due to debut
in December of '74, could fill Central Park,
not to mention the apartment.

Mary has been exploring Colorado Territory.

And Baby Arny has been vomiting.

Open/Close

My dear Mother Father & Guss,

Eunice almost cannot bear
to keep reading.

But She Does

Mary goes on to say
they abandoned a stateroom in Denver
for a cottage where the air thins
            for Arny's sake.

Six horses hauled their stagecoach
            up

up

the

mountain.

Mercifully, under a doctor's care,
Arny appears less *dreadfully*,

Mary writes.

His food
—oatmeal, beef tea, cow's milk—
did not assimilate
and gastric juice was weak.

Told me to stop his stimulents and milk
and he gave me receipts for two
preparations which to feed him.

He aimed to study
proper nourishment
and give as little
medicine
as possible.

Doctor's Orders

Mary, baby & big brother
must stay put for two weeks
while big John heads to Saint Louis.

I think
I am for the first time

on the right track.
It is really delightful here.

Affec'ly
Mary—

Sun

After enduring such a loss as Maud,

Eunice marvels at the optimism

still seeping from Mary's pencil,

like the filling of a boiling pie.

Speaking of Pie

When Mary returns home,
she thrusts herself
into a project—
an instruction manual
for aspiring socialites.

She tests hundreds of recipes,
bringing elegance to dishes like
Calf's Brains,
Tongue, with Mustard Pickle Sauce
& Fried Sweet-breads
(decidedly not breads).

To perfect a fluffy Chess-pie,
she ventures to the countryside
to hear the recipe expert's
explanations from her own lips.

Treatise

The secret of the pies
not becoming heavy,
the woman tells Mary,
is in cutting them
as soon as they are cooked,
and still hot.

If they are allowed to cool

they
will
fall.

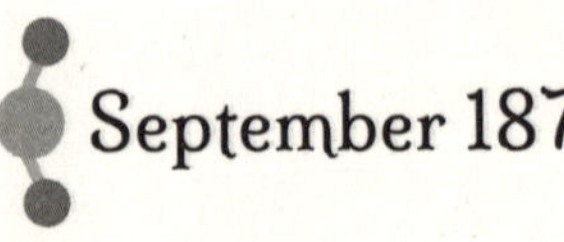

September 1874

Enrichment

When the baby fell ill in Colorado,
John caught gold-rush fever.
Now he tries to entice Eunice & Elisha
with his get-rich-quick scheme.

My Dear Judge, John writes to Elisha,
I have seen enough to satisfy me
that Mrs. Foote's chemical skill
will sooner find diamonds
in the separation of these ores
than in the chrystalization of carbon.

And I have further become convinced
that your mechanical powers
can be even better exerted
in this great field than even
in reapers and binders.

This is the field for Mrs. Foote.
The scenery is grand, the climate
charming, the mountains full of silver,
& the sands of the valley glistening
with gold. Nothing is needed but experts.

Hard-handed, brawny miners tear down
the granite mountains & follow
the quartz seams and roll down the ores—
but they know not how to work them
and are forced to sell them
for half of their value.

He who can
reduce them
cheaply is
the future
millionaire.

The Wrong Mountain

Eunice's
cheeks must flush
at her son-in-law's faith
in her skills. But the fever
doesn't catch. She has little desire
to strip minerals from mountains. She
wants to level the mountain between women

& equality.

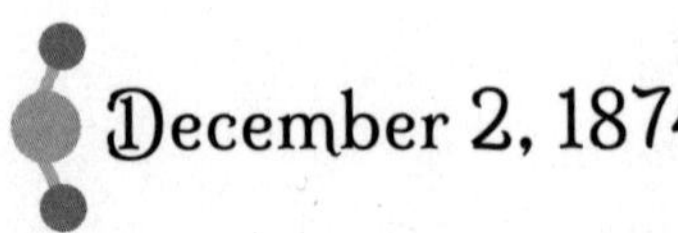

December 2, 1874

Gus Delivers

Finally,

a girl.

This Child

Frances.

Her name rests
on the shoulders
of those before her,
reflects the names

of her fallen brother,
her father,
her father's sister,
her father's mother.

Fanny's sweet cheeks,
her tiny squeaks,
her restless feet,

shoulder the hope
of generations
who still fight

for her future

for her voice

for her vote.

1874

Such As

In Missouri,

a woman who tried to vote,
who demanded that right
under the Fourteenth Amendment,
has been denied by the Supreme Court.

In Pennsylvania,

the new state constitution
strikes ~~*white freeman*~~ & expands
the vote to male citizens.

In Michigan,

a landslide majority of voters—
77 percent—say nay to the legislature's
proposed constitutional amendment
giving suffrage to women.

This Girl

Fanny will learn
how to talk,
how to walk,

how to draw,
how to write,
how to fight,

with one hand,
the other gripping
a torch she didn't ask for—

a flame no one asked for—
that singes those who fail
to carry it forward.

Has Eunice Failed?

She has cycled away
from women's crusader,
from published scientist,
from patented inventor,

to matriarch,
grandmother,
nurturer-in-chief.

Perhaps Eunice longs for her lab,
longs to renew old fingerprints
on the kaleidoscope of tools
that once ignited her soul.

Perhaps she swallows that feeling,
reminds herself that happiness is a choice.
At age fifty-five, she chooses to be at peace

with her version of now—
chooses to bask in the torch's warmth

from a distance.

Hot, Hot, Heat

All the rules Mary has laid out for women
hook the attention of editors—men—
at Harper & Brothers.

They offer her a contract to publish
Practical Cooking and Dinner Giving
in the year of our Lord, 1876.

She celebrates—
practically, of course—
with eminent grace & class.

Revolution

How much

can change

in a year?

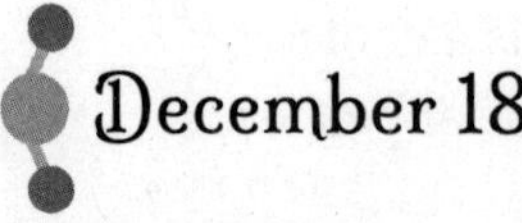

December 1875

Everything

My dear Mother, Father & Guss

Our poor little Arny

died

last night.

Gram

Eunice

is but a dot

hurtling through space

on a rock called Earth

staring at Mary's script.

Breathless

December,
cruelest of months.

Diphtheria,
cruelest of diseases.

The post,
cruelest of communications.

Mary

I never knew the child
to be in better spirits
than Sunday a week ago to day

he and Johnny just raced,
up & down the parlour
laughing & frolicking.

How?

Eunice's thoughts thaw,
at first a slow drip,
then a glacial gush.

How

could another
Henderson child
have perished?

How

could Arny have caught
the pestilence? Bad cheese?

Mary had done
what she could think to do—
what Eunice might have advised—
          castor oil,
                    rest,
                              fetch a doctor.

So how

did infection seize Arny's throat
& send the poor angel into spasms?

A Question Seizes Eunice

Johnny?
Little Johnny?
What about little Johnny?

Scanning

Frantic.
Scanning.
Frantic.
Scanning.

Until at last,

Mary's closing—

He seems perfectly well now but a cold,
but the doctor watches him like a hawk.

God grant
that he may
be spared us.

No

Grand- parents
should not live
to bury grand
children.

Entombed

Elisha buries his emotion

as any   man of the family
must.

He scrawls a note to the cemetery:

Hon. J. B. Henderson
                    man of his daughter

is authorized
to occupy
my tomb

        for the interment
        of his child–

for the untimely cycling
of carbon
back
to the Earth.

1876

Life Cycles

While little Johnny, age six,
          grows self-sufficient;

while big John busies himself
with President Grant's assignment
to take down the Saint Louis Whisky Ring;

Mary keeps the heat high
& throws herself into the Cause,

gaveling in as president of Missouri's
          Women's Suffrage Association,

          founding the Saint Louis School of Design
          to help women earn their own livelihood,

& eagerly awaits the moment
          she will hold her first book,
          the shred of what's left
          of before.

The Book Cycle

Weathering
                    of the author
                    during drafting.

Erosion
of ideas
in revision.

Deposition
of pages
on the printing press.

Compaction
over time,
so much time.

Crystallization
of dormant ideas
from another epoch.

Metamorphosis
into fossils
to be dug up
& distributed.

Published!

Eunice is thrilled to receive
her copy of Mary's book,

Practical Cooking and Dinner Giving

(*A Treatise Containing Practical Instructions
in Cooking; in the Combination and Serving
of Dishes; and in the Fashionable Modes
of Entertaining at Breakfast, Lunch,
and Dinner*).

But

in the long wake of a wake,
after all that's happened,
perhaps Eunice wonders:

Do these ideas still matter?

Mouthful of a Title

That's the nature
of womanhood:

walking the tightrope,
juggling every rule—

the only way for a woman
to earn enough trust

to attempt
her dreams.

But Dreams?

Dreams still matter.

That Eunice knows.

Rules

The book spans 396 dizzying pages
of how-tos
& what-not-tos
& recipes
for every-day living:

Do not put starch in the napkins,
as it renders them stiff and disagreeable. . . .

Place the soup-tureen (with soup that
has been brought to the boiling-point
just before serving) and the soup-plates
before the seat of the hostess. . . .

The waiter should wear a dress-coat,
white vest, black trousers, and white necktie;
the waiting-maid, a neat black alpaca
or a clean calico dress, with a white apron. . . .

Thirty Pages In

Mary's voice machetes through the morass:

I do not hesitate to condemn
and reject a custom in which I see
no good, but, on the contrary,
a temptation to positive evil.

There she is, Eunice's girl, writing her own rules.

Let it be remembered.

Catch

When thousands of dreams—
a nation's ambitions—
catch fire all at once,

do they sweep clean
the forest of progress
or burn themselves

out?

September 24, 1877

Fire Seeks a Path

Sparks float up a flue,
find wood on a copper roof,
turn orange metal white,
send solder dripping

& before you can yell its name,
the first-of-the-season blaze
meant to warm the west wing
of the cavernous US Patent Office

reduces a hulking edifice to a s h.

Incendiary

A crowd gathers,
stomachs rumbling
just before lunch,
gawking at the
Greek Revival
with a wig of smoke.

Fire   men
snake rubber hoses
up ladders through
top-story windows,
but flames chew
through copper roofing
to pine planks below.

Fire fights fire~~~~~~
    coal kindles boilers
        of horse-drawn steamers,
            clattering on cobblestone.
                Steam builds pressure,
                    pressure turns turbines,
                      steam turns gears,
                        faster, faster, faster.
                          Water shoots through
                              rubber hoses,
                                yet cannot
                                 reach
                                 the

                          ~~~~roof~~~~
                          ~~~~

Below That Roof

A relay of heroes crowd
curved stone stairs,
passing
models,
books,
papers
like baskets of rolls at mealtime—
thousands of models
for priceless inventions
saved in a seventy-five-foot-tall mound,

yet
a hundred tons
of rejected patent applications—
a hundred tons
of unrealized dreams
shelved below the roof

go *poof*

September 25, 1877

News Reports

The Philadelphia Inquirer:

> *A NATIONAL CALAMITY*
>
> ——————
>
> *FIRE IN THE PATENT OFFICE*
>
> ——————
>
> *Destruction of the Model Room*
>
> ——————
>
> *IRREPARABLE LOSS OF MODELS*
>
> ——————
>
> *Records Saved with Great Difficulty*
>
> ——————
>
> *WIDE FIELD FOR LITIGATION OPENED*
>
> ——————
>
> *Great Excitement of Patent Lawyers*
>
> ——————
>
> *Systematic Efforts to Save Valuables*

The Baltimore Sun:

> *At this time it is not possible to even conjecture the amount of the losses, but it is believed that at least 60,000 models of patents were burned. Such relics as the original draft of the Declaration of Independence, Gen. Washington's swords and camp equipage, &c., were saved.*

One Patent Lawyer & His Wife

What if

Eunice's & Elisha's patent applications
& their models could not be rescued?

What if

their life's work—submitted when
the patent office required one copy
of each application—disintegrated?

What if

the torch Eunice worked for,
the torch she fought for, blazed
out   of   control?

What then?

Smoldered

Their patents—
their patented hopes,
their hopeful dreams,
their dreamy future—

have weathered
the blaze

safely.

Thanks to the Fire Crews

The steam teams,
fire companies
from DC & Baltimore,
swooped in
& finally suppressed
the conflagration.

Perhaps
it occurs to Eunice
that one of the heroes—
one of the creators
of the machines
that snuffed the flames—

is Horace Silsby.

Silsby

Eunice's nemesis
now pumps out
more steam fire engines
than any other company
from his factory fortress
on a Seneca River island.

Perhaps
any lingering grudge
about stolen stove patents,
about years wasted in court,
washes away with the soot.

Excelsior.

The Rest

The end has come for Joseph Henry
at age eighty. Government shuts down
to pay respect to the man who discovered

the electric motor, revolutionized meteorology
& steered the Smithsonian from nothing
to greatness. Luminaries at his funeral

include the president, vice president,
cabinet members, Supreme Court justices
& military brass. But not the Footes.

They have been too busy to keep up
with the news. They know not
that their friend's light has gone out.

October 13, 1878

Electricity in Thunder-Storms

Elisha muses about the cause
of lightning in storms,
refuting a scientist's conclusion
that it is caused by friction.

It's from condensation, Elisha argues
before the Philosophical Society
of Washington, based on what
he's heard of Colorado's Pike's Peak,

& though he doesn't mention it,
based on Eunice's findings from years before.

He offers hypotheses
& conjecture for evidence.
The magazine edited by E. L. Youmans,
Popular Science Monthly, runs his article.

With his mountain of confidence,
Elisha is his own lightning rod.

Grounded

Elisha's staying
at the National Hotel
in Washington.

I occupy my old room in the upper story
at the National. The house is full but nearly
all strangers to me. Most of the old examiners
here called on me & manifest the same regard
that they had for me when I was Com'r.
They all have their grievances to relate. . . .

The office is filled with incompetent boys,
and a strong effort is being made
to abolish this office altogether.

Prof Henry is quite out of health
with a disease of the kidneys.
I have not yet seen him,

Elisha writes.

Flickering

Gas prices have soared
too high for light, people cry.

A man,

Thomas Edison,
phonograph inventor,
Wizard of Menlo Park,

has a solution
fit for the industrial age.

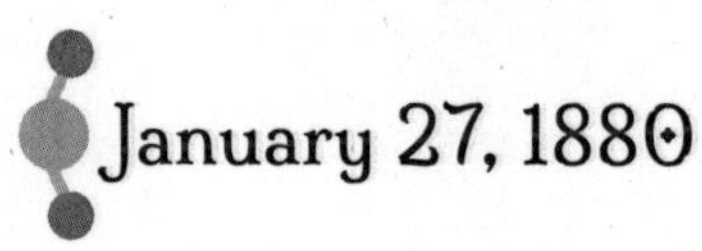

Edison's Lamp

Carbonized paper
two & a half inches long
threads like a horseshoe
in a glass globe exhausted of air.

Platinum clamps
on the ends of this carbonized paper
connect to platinum wires
crossed with copper wires
completing a circuit
with an electrical source.

Mr. Edison maintains,
with the utmost confidence,
that his electric light,
Patent No. 223,898,
is a complete success.

He insists
that he has solved every problem
that has hitherto puzzled scientists
in their efforts to make electricity
take the place of gas.

Eunice's Light

Carbon
stretched thin as thread,
trapped under glass,
exhausted,
clamped & wired,

yet somehow
still
lit.

July 26, 1881

Reaping Rewards

Elisha's most complicated invention,
his pièce de résistance,
the culmination
of his life's work, combining
mathematics & engineering
to sap sweat from harvest
for both human & horse
with moving
rods,
axels,
gears,
screws,
wheels
with teeth for gripping earth,

knives,
ribs,
stops,
bevels,
cylinders,
fingers,
arms for gathering,
carrying, lifting, throwing,
all meant to
slice,
propel,
drive
this substantially new machine
that cuts grain, binds bundles
with bands patented
eight years ago—
with support
from family,
from Eunice,
gifting time
to create
as well as space in their home,
now on West 39th Street
in New York City.

Another Elisha Foote
patent—
No. 244,876,
Machinery for Reaping
and Binding Grain—claimed
for the last time, for all time.

Be it known.

All Time

It promises infinite paths.

All paths vanish

once spent.

Spent

Regret
is a child
of unrealized paths.

The last time Eunice heard
from her beloved Mandy,
Amanda was upset about something.

Perhaps they stopped communicating.
Perhaps their relationship waned.
Perhaps they neglected to speak
of sisterly love.

One never knows
when a path will close
& electrify

regret.

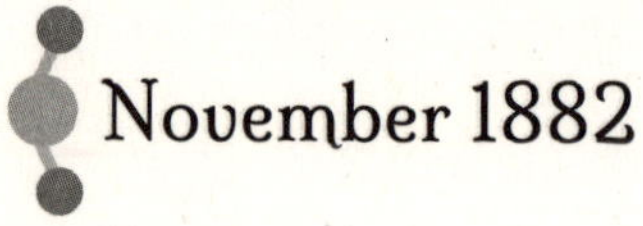

Mandy

Here lies
Miss Amanda Newton,
born fierce as a firework
on the Fourth of July,
resting in the power of eighty-two years—
working this blooming soil,
healthy until the moment
she fell down the stairs
& her collarbone
snapped.

She bought the farm?
Nay, she kept those 150 acres
  alive—with cows, swine & horses,
    with wheat, corn, oats, prize peas & hay,
      with a rotating cast of hired hands
                  & boarders
        & extended family.

And now that no Newton claims
this land, the homestead
will be sold.

Surprise

Because the last time Eunice
heard from Mandy she was upset,

when Eunice tears open
the letter from Gus, she's
not expecting this news:

Amanda has left Eunice
& siblings Darius & Adaline
each a hefty sum of $5,000—
almost nine years' wages
for the average American.

Gus writes:

I am very glad Aunty remembered you
equally with the rest. It shows
she really had no hard feelings
and her better judgment prevailed.

Relief blooms inside Eunice
like a bouquet sent from the grave.

Memories

Amanda will be remembered for

her
superior talent and judgment,
her
uncommon tact and resolution,
her
incessant toil.

In Saint Louis, Eunice folds up the letter
& busies herself with her remaining
days,
weeks,
months
& years
(God willing)
as Gram.

Journey

Elisha arrives in
Saint Louis, reconnecting
with his wife of forty-two years,
partner in love & invention, home-
making & lab-keeping, wondering
and travel, living, existing, aging,
exploring, ascending, creating,
moving, uprooting &
sharing this one
beautiful
life.

Rest

After tending to some business,
Elisha begins to feel unwell.

Nothing too unusual.
Nothing that won't clear

before Mary & John return.
Johnny's parents are en route

across the Atlantic Ocean,
fresh off the shores of France.

Mary marvels at the milk & cream
from sealed glass jars in the ice closet

served on the steamer's
final day, fresh as the first.

October 22, 1883

St. Louis Post-Dispatch Revised

Death of the Father-in-Law
of Gen. Henderson This Morning

Hon. Elisha Foote, father-in-law
of ex-United States Senator Henderson,

husband of inventor
Eunice Newton Foote,
father of suffragist
Mary Foote Henderson,

died
at the residence of his

daughter &

son-in-law, No. 3010 Pine street, at half-past
5 o'clock this morning, of heart disease.
The deceased gentleman was over
70 years of age.

Seventy-four, to be precise.

He arrived in St. Louis only a few days ago.
To a friend who met him on Saturday,
he expressed himself feeling only slightly
indisposed. During the administration
of President Andrew Johnson Mr. Foote
was Commissioner of Patents.
Gen. Henderson and wife

—Mary—

are not in St. Louis at present; at last
accounts they were in Paris. It is known
that Gen. Henderson expected to sail
for the United States on the Servia,
and that vessel was due at New York
this morning. Another daughter
of Mr. Foote's

—Augusta Foote Arnold—

is married and living in New York.

Home

After his last labored breaths from

heart disease and pulmonary congestion,

Elisha's body makes the long journey

from Saint Louis to its final resting place

in New York.

A Fitting Burial

Anyone who's anyone
makes their final sleep
among the winding paths
    & gentle hills
        & gilded monuments
    overlooking the harbor
at Brooklyn's Green-Wood Cemetery.

So much death in this place, yet so much life—
a half million annual visitors funnel through

the towering gothic arch with its haunting
sculptures of solemn bearded   men,

take jaunty carriage tours
—clip-clop, clip-clop—
past the tombs of celebrities such as

Samuel Morse,
inventor of Morse code,

Horace Greeley,
New York Tribune founder,

Henry Steinway,
pianoforte magnate,

& now
the sandstone tomb
of Judge Elisha Foote.

Together

Eunice has followed
Elisha one last time.

She lives in New York now
& can ferry across the harbor

to this scenic sanctuary, can crunch
the yellowed leaves, can reflect

on a lifetime enshrined
on this rectangle

of carbonaceous earth,
where she too will rest

some day,
some week,
some month,
some year,

in a future that feels
close enough
to touch.

Past, Present, Future

Eunice couldn't have asked
for a better partner than Elisha.

She cannot imagine a man
who has supported women more.

She will never experience
greater respect, unity, love.

1885

Driven

Mary flings herself into a research vortex,
scouring journals & articles, quizzing doctors,
seeking recipes to learn how to stay alive
by eating the right foods.

Of course, she knows
health is not that simple,
but after so many deaths,
one looks for things
to blame.

Diet for the Sick

While Mary's first book grazed
the skin of her life's purpose,
Diet for the Sick cuts to the heart.

Gold-leaf lines course the cover
like the fingerprints of her losses.

"Man
kills himself, rather than dies,"

reads the epigraph, followed by
229 pages of recipes based on
cutting-edge 1880s science
to nurse the *invalid* to health.

Page 76, *Diphtheria*
haunted by the ghost of Arny.

Page 102, *Beef Tea for Travelling*
echoes the trip where Arny convalesced
in the Rocky Mountains.

Page 214, *Food for Infants*
alludes to the anguish of burying
a cherub like Maud.

To Hold On to Health

Say no to
lavish ice water!
coffee & tea!
spices (a gateway to alcoholism)!
too much salt!
processed foods!

Say yes to
Hard work like shoveling sand, sweeping,
or the new patented gymnastics machines
(if one doesn't mind waving one's arms
with no practical purpose).

When health is everything,
life itself, how little it is guarded!
Mary writes.
How little appreciated,
except when lost!

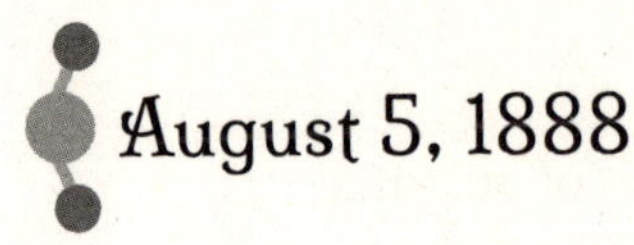

August 5, 1888

Driving

The three-wheeled Motorwagen
with an internal combustion engine
fueled by gasoline—
byproduct of kerosene,
byproduct of dead dinosaurs,
byproduct of not paying attention
to a woman's breath-altering
discovery—

takes its maiden cruise.
A woman named Bertha Benz pilots
her husband's experimental machine,
bumping across the German countryside
with her two teen sons to visit her mother.

What if
Bertha had known?

What if
she had listened to Eunice?

Would she
have pushed this invention
to change the world?

Reflections

In her lifetime, Eunice has
found carbon dioxide's warming properties,
discovered lightning's cause,
eroded barriers for women in science,
claimed two patents in her name,
raised two children,
loved her grandchildren,
achieved 42 years of egalitarian marriage.

In her lifetime, her country has
built the Erie Canal,
connected the coasts by rail,
removed Native people,
fought a civil war,
emancipated enslaved people,
carved out women's rights,
borne the oil industry,
invented convenience
through kerosene
& vulcanized rubber
& the telegraph
& the telephone
& the electric light
& the automobile.

Of all the noteworthy moments,
which ones flash at the end?

Signing the Declaration of Sentiments?
Sharing the warming powers of CO_2?
Earning less recognition than she deserved?

Or

Does Eunice focus on tender times?

Digging potatoes with a grandson?
Nurturing her daughters as mothers?
Dining on the river with her beloved?

September 29, 1888

In the End

In Berkshire County, Massachusetts,
not far from Elisha's childhood home,

Eunice closes her eyes
to all she's seen—

the developments,
the inventions,
the progress for women,

the regret.

After coughing so long,
lungs begging for clear air,

she succumbs at age sixty-nine
to *bronchitis.*

In the And

Her spirit

breaks free.

Excelsior.

One Hundred Twenty-Two Years Later, 2010

Seismic

Somewhere in Oklahoma, a state formed
almost two decades after Eunice left this life,
a retired petroleum geologist scans
his basement library of antique science books.

Hungry for an origin story,
Raymond Sorenson traces
the worn covers until a title calls to him:
The Annual of Scientific Discovery.

At the crack of the spine,
long-dormant plates shift.
Aftershocks ripple, a century
& a half past their genesis.

Ray lands on page 159,
where the epicenter explodes
his understanding of climate science
& carbon dioxide:

An atmosphere of that gas
  would give to our earth
    a much higher temperature.

Ray checks the date of the report—1856.
        How   can   this   be?

John Tyndall discovered the greenhouse effect
in—Ray checks his notes—1859 . . . right?
John Tyndall, *the father of climate science*?

Ray claws through memory,
his own & the collective, searching
for mention of the soothsayer responsible
for the 1856 declaration he just read:
                    Mrs. Eunice Foote.

Finding none, he exhales & writes an article,
a puff of carbon fogging the glass ceiling
so the world can trace her name.

Eunice.
Newton.
Foote.

Be it known.

A Note from the Author

Although the many cited facts in this book are true and the general trajectory of Eunice's life is accurate, this is a work of fiction because of some of the connections I made. I made occasional assumptions: For example, in the poem series "September 1839, Property," "Gossip," and "Pressing," I portray Eunice as reading a news report about Troy Female Seminary director Emma Willard. I cannot know whether Eunice read about Emma's divorce, which was covered in the papers. But the news was relevant to Eunice's narrative arc, and I decided to include it. Likewise, in the poem series portraying Eunice's trip on the SS *Great Eastern*, I was unable to ascertain whether she and Mary received their passports in time and actually made the journey. The information about Elisha writing to the secretary of state and details about the people they were meant to travel with are true. There are other examples like this throughout the text. I did everything in my power to stay close to the facts as I told Eunice's story.

More About Eunice

Eunice Newton Foote is now recognized as the first person to discover the warming properties of carbon dioxide, which she called "carbonic acid gas" at the time of her 1856 experiments. She concluded, "An atmosphere of that gas would give to our earth a high temperature."

Through her work, she intended to explain the warmer climes of earlier epochs when dinosaurs roamed. What she didn't foresee was a warmer atmosphere due to human activity—resulting from many of the inventions that evolved during her lifetime.

She recognized that if she wanted to make an impact as a scientist, she must fight for her right to be one. She became involved in the nascent women's rights movement by attending the 1848 Seneca Falls Convention. She signed and helped publicize the revolutionary women's rights manifesto modeled after the Declaration of Independence.

The more I learned about Eunice and the intersections of her life's work, the more my journalistic instinct drove me to piece together her important story. She was related to Sir Isaac Newton—ancestry records confirm it. She attended the first state-funded secondary school in the United States to teach science to girls. She signed the Declaration of Sentiments and was involved with Elizabeth Cady Stanton's family. Her husband ascended to become the most powerful man in the US Patent Office, and she earned several patents during a time when few other women could claim the same. The Footes' engagement with invention occurred during a time when many of the patents being awarded furthered the cause of industrialization and use of fossil fuels. Those same fuels, when burned, released the very "carbonic acid gas" Eunice had warned about.

Like many women of her time, Eunice's main focus was her family. Still, she managed to make incredible strides in science and invention, and she passed those values on to her children. Eunice's older daughter, Mary Foote Henderson, rose to wealth and influence after she married Senator John Brooks Henderson, who earned his spot in history by introducing the Thirteenth Amendment to abolish slavery and by preventing Abraham Lincoln's successor, Andrew Johnson, from being removed from office. Mary carried the women's rights torch and forged opportunities for women. Before her death in 1931, she saw what her mother only imagined: suffrage via the Nineteenth Amendment in 1920. Mary, however, was not without controversy: late in life she donated two million dollars to the Race Betterment Foundation, which promoted eugenics and "racial hygiene." Racism like Mary's was atrocious during her time, as is it now.

Eunice's younger daughter, Augusta Foote Arnold, became a scientist and author. In addition to two cookbooks published after Eunice's death under a pseudonym, her tome *The Sea-beach at Ebb Tide: A Guide to the Study of the Seaweeds and the Lower Animal Life Found between Tide-Marks* (1901) later appeared as a reference in the writing of influential conservationist Rachel Carson.

Despite Eunice's monumental discovery about the potential for a warming Earth, for more than 150 years the world insisted that an Irish male scientist, John Tyndall, had founded the field of climate science. His experiments on the warming properties of various gases occurred three years after Eunice's were published in *The American Journal of Science and Arts*. Some argue that because Tyndall's work (on an unrelated topic, colorblindness) appeared in the same publication, he must have read Eunice's report before he claimed to the Royal Society in 1859 that "Nothing, so far as I am aware, has been published on the transmission of radiant heat through gaseous bodies."

It wasn't until 2010 that a retired petroleum geologist named Ray Sorenson lucked upon a report about Eunice's 1856 write-up, "Circumstances Affecting the Heat of the Sun's Rays," and decided to share her discovery. His article about Eunice appeared in a 2011 issue of *Search and Discovery*, "an online journal for geoscientists working in energy fields." Since then, word of Eunice's accomplishment has spread only at a trickle, mostly online. I first learned her name in 2021 and became fascinated with the what-ifs:

What if the world had listened to Eunice?

What if the world had trusted women as intellectual, competent beings during her lifetime?

What if women could have voted all along?

What if environmental impact had been considered in patent applications?

What if humans hadn't forged ahead with fossil fuels?

What if we weren't in a climate crisis today?

May we all fearlessly chase our dreams in the way Eunice did. May we all love selflessly, as she did. And may we all listen to the scientists who decry the burning of fossil fuels.

Eunice's

Bolded names are referenced in this book.

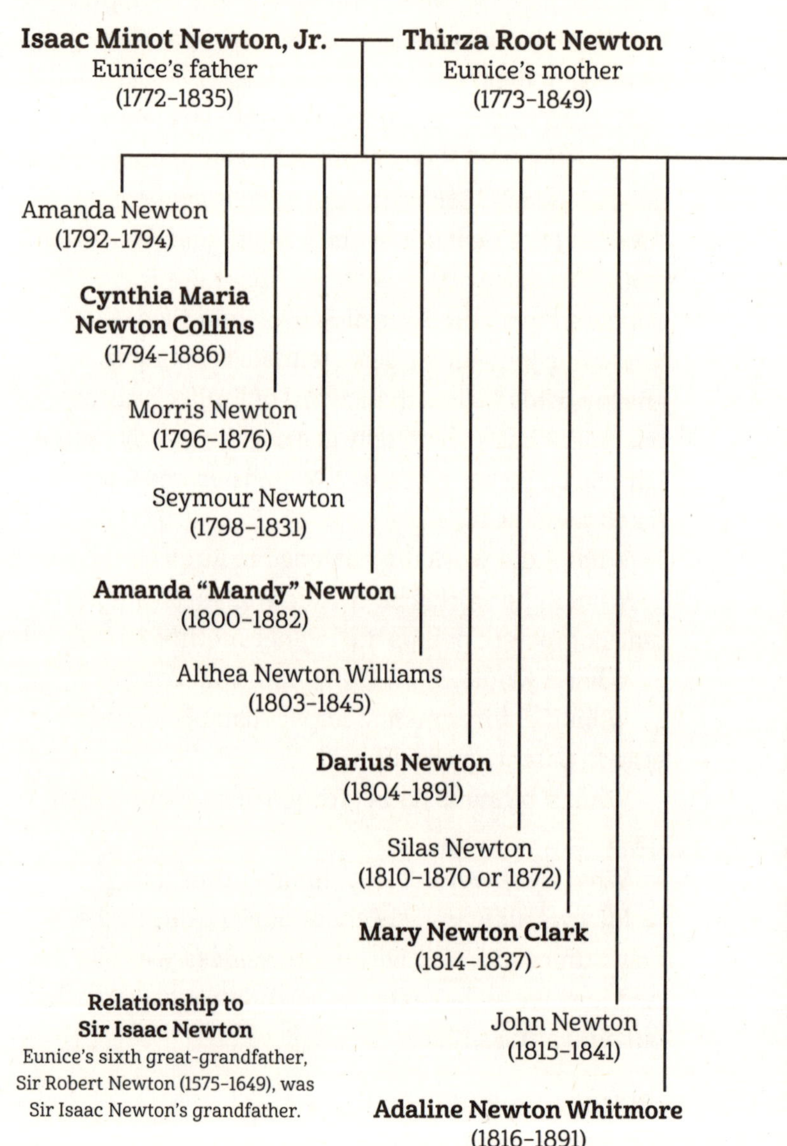

Relationship to Sir Isaac Newton

Eunice's sixth great-grandfather, Sir Robert Newton (1575–1649), was Sir Isaac Newton's grandfather.

Family Tree

2 of 13 siblings
Eunice Newton Foote (1819–1888)
Elisha Foote (1809–1883)
Henry Rutger Foote (1824–1899)
Mary Foote Henderson (1842–1931)
John Brooks Henderson (1826–1913)
John "Johnny" Brooks Henderson, Jr. (1870–1923)
Maud Henderson (circa November 1871–August 1872)
Francis Arnold "Arny" Henderson (1873–1875)
Francis "Frank" Benjamin Arnold (1844–1906)
Augusta "Gus" Foote Arnold (1844–1904)
Benjamin "Benny" Foote Arnold (1870–1896)
Francis Arnold, Jr. (April 1872–September 1872)
Henry Newton Arnold (1873–1939)
Frances "Fanny" Arnold (1874–1975)

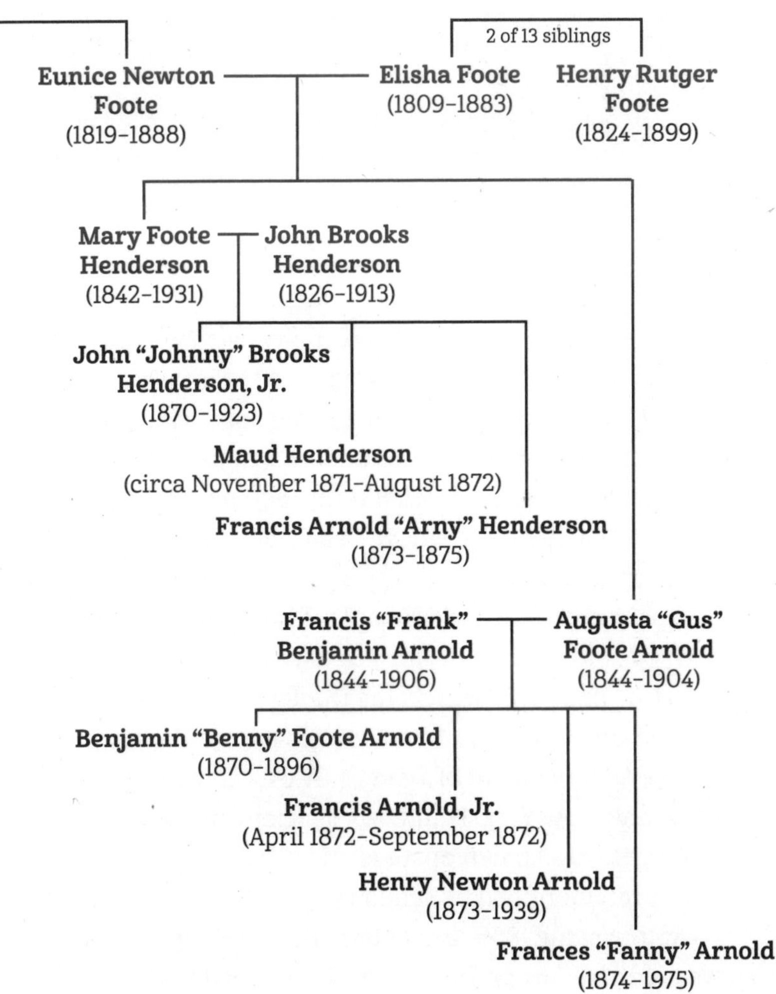

Could This Be Eunice?

Detail from Matteson's Dudley Observatory Dedication, August 28, 1856.

Because of Eunice's forced invisibility as a woman, as of this writing in 2025, nobody is sure what she looked like. The world has not confirmed a portrait of her despite the efforts of hundreds, or perhaps thousands, of researchers. But through my research, I have isolated what I believe to be a portrait of Eunice and Elisha.

A painting by the artist Tompkins Harrison Matteson entitled *Dudley Observatory Dedication, August 28, 1856* (see page 156) depicts the Dudley Observatory dedication at the end of the 1856 American Academy for the Advancement of Science (AAAS) convention in Albany, New York. Hundreds of scientists and associates are shown onstage and off; only some have been identified by scholars.

At that same 1856 convention, Eunice's paper was read by Joseph Henry, and Elisha read his own paper. I have to think Elisha and Eunice would have attended the observatory dedication, although they've not been identified in this painting.

As I scanned the painting for prominently placed but unidentified women, I stopped at a couple onstage just behind the speaker's arm. The man bears a striking resemblance to Elisha Foote's official patent office photograph (see page 154), taken in the 1860s when Elisha was in his late fifties. The man in the Matteson painting appears to be wearing glasses, and his build, beard, hairline, and facial features all seem to match Elisha's. I believe the image in this painting looks like Elisha would have appeared in 1856 at age forty-seven. What's more, the woman seated next to this Elisha look-alike matches Eunice's physical description from her passport application. (See the poem "Superficial" on page 116.)

The couple is seated prominently, much like the first female AAAS member, Maria Mitchell, and her husband, who have been identified by scholars in this same painting. Elisha was voted in as a member at this convention, so it follows that he would appear onstage. It's also uncanny how much Eunice and Elisha's daughter Mary looked like the woman here (see page 155).

Viewing the painting in person at the Albany Institute of History & Art only deepened my suspicions.

I showed the painting to experts. Leif HerrGesell, a historian for Ontario County, New York, and former director of the East Bloomfield Historical Society in Eunice's hometown, agreed the couple could be the Footes. "I think the gentleman is almost a dead ringer for Elisha," he wrote in his email reply.

"I would hazard that their placement in the image is also an indication of their importance. All of the other women are either too old or placed farther back suggesting less importance, and the woman appears to be the correct age and the gentleman just a little older than her. Images of this type were always very carefully laid out to feature people according to rank and importance. . . . I would even publicly state that you have isolated the best possibility! I think you found her."

Sam McKenzie, a historian formerly of Saratoga Springs, New York, has studied Eunice's life. "The man sitting next to her bears a strong resemblance to presumably properly identified later photographs of Elisha," he wrote to me. "Well done, you!!! I think this is called . . . a scoop."

Despite the experts' opinions, confirming that it is Eunice in the painting will be difficult. Some historians take a more conservative approach and wait for incontrovertible evidence, which we do not have at the time of this writing in 2025. If this is Eunice, it would be the first image identified of her worldwide—an assertion I do not take lightly. With or without Eunice's likeness, her story—her legacy—can inspire future generations.

Mary's Full Letter

Excerpts of an 1890 letter penned by Mary Foote Henderson appear at the beginning of each part of this book. The letter is an homage to the famous suffragist Susan B. Anthony, who knew Eunice and Mary. Here is Mary's letter in its entirety.

Washington
Feb 16, 1890

Mrs. Sewall, Chairman of the Banquet given to Mrs. Anthony.

Dear Madam,

From childhood I have respected and admired Susan B. Anthony.

Among my early teachings, not the least valuable, was a proper regard for the little band of women, who forty years ago inaugurated the movement, whose blessings are now recognized and confessed by all, or nearly all. These women were then unpopular. Although their lives were so blameless as to defy calamity, the weapons of ridicule were substituted, and used against them with relentless power.

They fully realized the work before them. Possessed of that material of which martyrs are made, they dispised [sic] *the counsels of fear, and worked with untiring zeal, for the purpose of giving women every fair chance in life, ~~for them to~~ that they might become instead of possible burdens or indulgent men, comparatively independent and useful citizens.*

Truly, "the little heaven" they placed in society has finally changed the whole face of our civilization.

Women now hold all places in which intelligence, fidelity, honesty and sobriety are the requisites of gratification. They are the best of salesmen, book-keepers, telegraph operators. They are among our best orators, and do give work in nearly all the trades and professions. Beside this, most acceptably do they administer the duties of public office when entrusted to them.

Mrs. Anthony and ~~Mrs. Stanton~~ her brave associates of earlier years may not live to realize the full fruition of all their hopes, but much is already accomplished, and the rest will come in due time.

I wish together with the great army of women that now appreciates Mrs. Anthony's work to express my great sincere admiration for that noble, capable and brave woman.

I am

Very Truly Yours
Mary F. Henderson

Source Notes

Direct quotations from sources are italicized in the text. Below are the sources listed by part and then by poem title.

Part One

Opener: "They fully realized . . . in life": M. F. Henderson, Letter to May Wright Sewall, pp. 2–3.

Up: "Excelsior!": New York Department of State.

Scientist: "man of science": Bergland, pp. 146–147; "Scientist": Ross, p. 71.

Gravity: "satellites of men" and "primary existences": Willard, p. 15.

Progress: "marvel": Robb.

Kaleidoscope: "ladies": Ricketts, p. 61.

Gossip: "distinguished . . . property": *Southern Argus*, p. 3.

Regulating Heat II: "Regulating . . . Stoves," "when . . . degree," "uniform," "a more . . . control," "the requisite intensity," and "Be it known": Elisha Foote, *Regulating*, pp. 1, 3, 5.

Part Two

Opener: "They might become . . . citizens": M. F. Henderson, Letter to May Wright Sewall, p. 3.

Living History: "for the increase . . . men" and "Smithsonian Institution": Smithsonian Institution Archives.

Eunice Wonders: "Smithsonian Institution": Ibid.

Real: "Well known . . . in use": *Seneca County Courier.*

Her Time: "a patent . . . invention" and "as if . . . unmarried": *Brooklyn Daily Eagle*, p. 2.

Catching Wind: "self-registering apparatus" and "which perhaps . . . Institution": Elisha Foote, Letter to Joseph Henry.

Sunny Skies: "500 . . . public": *Ithaca Daily Chronicle.*

Married Women's Property Act: "as if . . . female": *Brooklyn Daily Eagle and Kings County Democrat*, "Chap. 200," p. 4; "obey": Griffith, p. 44.

Three Days Later: "do . . . follows," "one hundred and forty-one miles," "sixty-seven miles," "one hundred and fifty-six miles," and "the sum . . . appropriated": *Brooklyn Daily Eagle and Kings County Democrat*, "Chap. 213," p. 4.

Intersections: "JUST RECEIVED . . . market" and "patent": *Nunda (New York) Democrat*, and "Race Regulator": Race, p. 3.

Call for Convention: "to discuss . . . Woman": National Park Service, "Participants."

And Yet: "Resolved . . . equal," "demand," "declare," and "secure": Lewis.

Volunteers: "Lucretia Mott . . . Newton Foote": Library of Congress; "Misconception . . . ridicule": National Park Service, "Declaration of Sentiments."

Currents: "Right Is . . . No Color": *North Star*, p. 1; "Declaration of Sentiments," "embracing . . . country," "state . . . Legislatures," "the pulpit," and "the press": Dick, p. 10.

Science: "The work . . . age": *Proceedings of the American Association for the Advancement of Science*, p. 67.

Wavelengths: "Married women . . . property," "If women . . . souls," and "then they . . . any laws": Heinemann, p. 21.

Apparently Not Seward: "Infringe . . . afterwards," "Not substantially different," and "Affirmed": Justia.

The Sun: "that nearly . . . sun": Henry, p. 38.

She Knows: "carboniferous": Ortiz and Jackson, p. 76.

Until, Again: "The receiver . . . cooling": Eunice Foote, "Circumstances," p. 383.

Convention II: "On the Heat . . . Rays": Wells, p. 159.

Today: "Mrs. Elisha Foote": Wells, p. 159; "Circumstances . . . rays": Eunice Foote, "Circumstances," p. 382.

Professor Joseph Henry: "Mrs. Elisha Foote," "science was . . . the true," and "An atmosphere . . . temperature": Wells, p. 159.

Reception: "Lastly . . . your readers": *New York Daily Herald*, "The Scientific Congress," p. 8.

Plucked: "On the Heat in the Sun's Rays": Elisha Foote, "On the Heat," p. 167.

Back to the Lab: "Becquerel . . . Humboldt": Mrs. Eunice Foote, "On a New Source," p. 124; "electrical excitation" and "may be . . . air": *New York Daily Times*, p. 2.

Convention III: "ladies engage," "On a New Source of Electrical Excitation," and "Mrs. Elisha Foote": Ibid.

But, in November: "I . . . excitation": Mrs. Eunice Foote, "On a New Source," p. 123.

A Divided Supreme Court Decides: "The award . . . affirmed": Casetext.

Partial Victory: "electrical excitations" and "Mrs. Elisha Foote": Mrs. Elisha Foote, p. 239.

Drake's Folly: "Drake's Folly": McCully, p. 7.

Whose Story?: "folly": Ibid.

Soul: "Eunice N. Foote" and "Mary Newton Foote": E. N. Foote, p. 1.

Eunice Writes: "My improvement . . . manner," "cut and pared . . . shape," and "Be it known": Ibid., p. 2.

Answering the Call: "to all loyal citizens": United States Senate.

Superficial: "full forehead," "ordinary nose," "rather large," "small/ ordinary," and "sallow": Ancestry.com, "U.S., Passport."

Elisha Writes to Seward: "Dear Sir . . . Elisha Foote": Elisha Foote, Letter to William H. Seward.

Behold: "deficient in strength," "cheap and inferior," "I have . . . defect," "Eunice N. Foote," and "A. Newton": Eunice N. Foote, *Improvement*, pp. 2–3.

Ice: "Be it known": Elisha Foote, *Skates*, p. 2.

Interlude: "A gentleman . . . character": US Patent and Trademark Office.

A Wrap: "raw . . . consumed": *Leavenworth Daily Conservative*, p. 1.

Fluid: "carbon dioxide": Merriam-Webster Dictionary.

Epicenter: "Burning Hydrocarbon" and "By such . . . value": H. R. Foote, *Improved*, pp. 1–3; "first" and "by flames . . . oil": H. R. Foote, *Description*, p. 53; "carbonaceous," "productive," "powerful," and "heat": Ibid., p. 2.

Black Hole: "crude oil . . . a fuel," "It is perfectly safe," "economical," and "More steam . . . money": Ibid., p. 52.

What Next?: "supplied only . . . consumed": Ibid.; "firemen" and "coal-passers": Ibid., p. 45; "science was . . . nature," "if a success . . . itself," and "man . . . world": Ibid., p. 53; "increase the intensity" and "spread the flame": H. R. Foote, *Improved*, p. 2.

The President Impeached: "Hear ye! Hear ye!": Riddick and Dove, p. 19; "wretched man": Trickey.

Mary's Wedding: "daughter of . . . Foote" and "Mrs. Foote": *New York Daily Herald*, "Fashionable," p. 8.

Patent Office: "museum of curiosities": Robertson, p. 33.

Honest: "PRINCIPLE, NOT . . . LESS": *The Revolution*, masthead.

Illumination: "No doubt . . . women," "men had the money," and "loved notoriety": Stanton, p. 2.

Part Three

Opener: "These women . . . power": M. F. Henderson, Letter to May Wright Sewall, p. 2.

Frankly: "Frank": A. F. Arnold, Letter to Mother, p. 1.

Transcontinental Railroad: "Almost ready . . . offered," "The spike . . . presented," "The signal . . . blows," and "Done": *New York Times*, p. 1.

Eunice's Beginning: "I wonder . . . die": Eunice Newton Foote, Letter to M. F. Henderson, "NY Jan 1st," p. 3; "Johnny": Eunice Newton Foote, Letter to M. F. Henderson, "Dec 7th," p. 2.

So Much Has Changed: "Aunty": Eunice Newton Foote, Letter to M. F. Henderson, "NY Jan 1st," p. 3.

Eunice Frets: "I wish you . . . it to me," "Aff—," and "Mother": Eunice Newton Foote, Letter to M. F. Henderson, "East Bloomfield Nov 7th," p. 1.

Nesting: "Mother": Ibid.

Letter: "Happy . . . child," "Benjamin is . . . makes a fus," "Guss is . . . a few days," "The day," and "is charming": Ibid., pp. 1–3.

Babies: "the finest . . . was": Elisha Foote, Letter to M. F. Henderson, p. 1.

But Soon: "I will . . . soon": Eunice Newton Foote, Letter to M. F. Henderson, "Washington Sunday 19th," p. 3.

Safe in Saint Louis: "Gutted": *Evansville Journal*, p. 3.

Babymoon: "his Excellency": J. B. Henderson, Letter to M. F. Henderson, "Tuesday."

Maud: "Maud": Green-Wood Cemetery Archives, "Henderson, Maud"; "might in battle": BabyCenter.

John Writes to Mary: "I think . . . stand it?": J. B. Henderson, Letter to M. F. Henderson, "Washington Thursday 11 p.m."

Back to East Bloomfield: "Improvement in Driers" and "whether for . . . purpose": Foote and Smith, pp. 1 & 5.

It's a Boy!: "Francis Arnold, Junior": Ancestry.com, "Paris."

Changes: "My Dear Wifey," "she is not . . . happiness," and "Hubby": J. B. Henderson, Letter to M. F. Henderson, "Washington July 17 1872."

France: "The best baby . . . seen," "the loveliest . . . ever seen," "a most . . . leave," and "châteaux": A. F. Arnold, Letter to Elisha Foote, pp. 1–10; "quite a Frenchman": A. F. Arnold, Letter to M. F. Henderson, p. 2.

Labor: "The people . . . men," "the former . . . bent double," and "Mary . . . Gus—": A. F. Arnold, Letter to Elisha Foote, pp. 7–10.

Oh Là Là: "I hear . . . weather": A. F. Arnold, Letter to M. F. Henderson, p. 8.

Letters to Long Branch: "I am uneasy . . . from you" and "If either child . . . at once": J. B. Henderson, Letters to M. F. Henderson, "Monday 13 Aug."; "My Dear Wifey," "What is . . . write?" "I am so . . . children," and "Affectionately . . . Papa": J. B. Henderson, Letters to M. F. Henderson, "St. Louis Mo Saturday 17"; "I am so . . . save her," "Your mother . . . telegraphed him," "I have . . . speak," and "on Thursday . . . Saturday—": J. B. Henderson, Letters to M. F. Henderson, "Monday morning Aug 19."

Maudlin: "congestion of the brain": Green-Wood Cemetery Archives, "Henderson, Maud."

From Maud (Part XVIII): "There is . . . none" and "My own . . . farewell": Tennyson.

It So Happens: "F O O T E": Saratoga.

Harvest II: "It makes . . . hard": A. F. Arnold, Letter to Elisha Foote, p. 7.

Healing & Growing: "Poor little . . . pauper," "brought in . . . girl," "the French . . . with her," "I will . . . & all," and "Love to Auntie . . . Mary": M. F. Henderson, Letter to Mother, pp. 1–4.

And by the Way: "I received . . . for you": Ibid., p. 2.

President Grant!: "As far . . . easily enough": Ibid., pp. 2–3.

Mary Will Not Give Up: "There is the . . . anywhere" and "All of the . . . laugh": Ibid., p. 3.

Meanwhile, Beyond the Sea: "Paris is . . . place": A. F. Arnold, Letter to Elisha Foote, p. 6.

Patently: "Be it known": Elisha Foote, *Grain-Bands*, p. 2.

Reunited: "I think . . . in NY": A. F. Arnold, Letter to M. F. Henderson, p. 3.

Witness: "M. F. Henderson" and "It becomes heated . . . produced": Elisha Foote, *Gas-Burners*, pp. 1–3.

Sweet Joy, It's a Boy: "sickness": E. N. Foote, Letter to M. F. Henderson, "East Bloomfield Nov 7th"; "Henry Newton Arnold": Gurganus.

Mary's Turn: "Francis Arnold": Ancestry.com, "Francis A Henderson"; "Arny": M. F. Henderson, Letter to Mother, Father, and Guss, "Sunday," p. 1.

Open/Close: "My dear . . . & Guss": M. F. Henderson, Letter to Mother, Father & Guss, "Idaho Springs Monday," p. 1.

But She Does: "dreadfully" and "His food . . . as possible": Ibid., pp. 1–2.

Doctor's Orders: "I think . . . right track": Ibid., p. 1; "It is . . . here" and "Affec'ly Mary—": M. F. Henderson, Letter to Mother, Father & Guss, "Idaho Springs Col Sunday," p. 3.

Speaking of Pie: "Calf's Brains," "Tongue, with Mustard Pickle Sauce," "Fried Sweet-breads," "Chess-pie," and "explanations . . . lips": M. F. Henderson, *Practical Cooking*, pp. 151, 145, 152 & 240.

Treatise: “The secret . . . cutting them,” “as soon . . . still hot,” “if they . . . cool,” and “they will fall”: Ibid., p. 241.

Enrichment: “My dear Judge,” “I have seen . . . binders,” and “This is the field . . . millionaire”: J. B. Henderson, Letter to Elisha Foote, pp. 1, 4, & 6–7.

Such As: “~~white freeman~~” and “male citizens”: Americans All.

Everything: “My dear Mother . . . night”: M. F. Henderson, Letter to Mother, Father & Guss, “Sunday,” p. 1.

Mary: “I never knew . . . frolicking”: Ibid., p. 3.

Scanning: “He seems . . . spared us”: Ibid., p. 7.

Entombed: “Hon. J. B. . . . his child—”: Elisha Foote, Letter to Green-Wood Cemetery, p. 1.

Life Cycles: “earn . . . livelihood”: Hamlin.

Published!: “Practical Cooking . . . Giving” and “A Treatise . . . and Dinner”: M. F. Henderson, *Practical Cooking*, p. 5.

Rules: “Do not put . . . disagreeable,” “Place the . . . hostess,” and “The waiter . . . apron”: Ibid., pp. 14, 16 & 23.

Thirty Pages In: “I do not . . . evil” and “Let . . . remembered”: Ibid., pp. 30 & 17.

News Reports: “A NATIONAL . . . Lawyers” and “Systematic . . . Valuables”: *Philadelphia Inquirer*, p. 1; “At this time . . . were saved”: *Baltimore Sun*, p. 1.

Electricity in Thunder-Storms: “Electricity in Thunder-Storms”: Elisha Foote, “Electricity.”

Grounded: “I occupy . . . relate” and “The office . . . seen him”: Elisha Foote, Letter to E. N. Foote, pp. 2–3.

Edison’s Lamp: “Carbonized paper,” “exhausted,” “the ends . . . paper,” “Mr. Edison maintains . . . success,” and “He insists . . . of gas”: *Evening Star*, p. 2.

Reaping Rewards: “substantially new machine,” “Machinery for Reaping and Binding Grain,” and “Be it known”: Elisha Foote, *Machinery*, p. 9.

Surprise: “I am very . . . prevailed”: A. F. Arnold, Letter to E. N. Foote, p. 1.

Memories: “superior . . . judgment,” “uncommon . . . resolution,” and “incessant toil”: Cott, p. 5.

St. Louis Post-Dispatch Revised: “Death of the . . . living in New York”: *St. Louis Post-Dispatch*, p. 5.

Home: “heart disease . . . congestion”: Ancestry.com, “Elisha Foote.”

Diet for the Sick: “Man kills . . . dies,” “Diphtheria,” “Beef Tea . . . Travelling,” and “Food for Infants”: M. F. Henderson, *Diet*, title page, 76, 102 & 214.

To Hold On to Health: "lavish": Ibid., p. 6; "with no . . . purpose" and "When health . . . lost!": Ibid., p. 56.

Driving: "Motorwagen": The Automotive Hall of Fame.

In the End: "bronchitis": Green-Wood Cemetery Archives, "Foote, Eunice N."

Seismic: "An atmosphere . . . temperature" and "Mrs. Eunice Foote": Wells, p. 159; "father of climate science": Hume.

More About Eunice

"carbonic acid gas": Eunice Foote, "Circumstances," p. 383.

"An atmosphere . . . temperature": Wells, p. 159.

"racial hygiene": Wellcome Collection, p. 146.

"Nothing, so far . . . bodies": Tyndall, p. 37.

"an online journal . . . fields": Search and Discovery.

Could This Be Eunice?

"I think the gentleman . . . found her": HerrGesell.

"The man sitting . . . a scoop": McKenzie.

Mary's Full Letter

"Washington Feb 16 . . . Mary F. Henderson": M. F. Henderson, Letter to May Wright Sewall, pp. 1–5.

Source Notes Bibliography

The sources below are those that are quoted directly.

Americans All. "Timeline of the Women's Suffrage Movement: 1874–1887." americansall.org/legacy-story-group/timeline-womens-suffrage-movement-1874-1887.

Ancestry.com. "Missouri, U.S., Death Records, 1850–1931 for Elisha Foote." Provo, UT, 2008. Original data: Missouri Death Records. Jefferson City, MO: Missouri State Archives.

———. "Missouri, U.S., Death Records, 1850–1931 for Francis A Henderson." Provo, UT, 2008. Original data: Missouri Death Records. Jefferson City, MO: Missouri State Archives.

———. "Paris, France, Births, Marriages, and Deaths, 1680–1930 for Francis Arnold." Lehi, UT, 2021. Original data: Archives de Paris. Paris, France: Actes de naissance, de mariage et de décès.

———. "U.S., Passport Applications, 1795–1925 for Eunice N. Foote." National Archives and Records Administration (NARA); Washington D.C.; Roll #108, Volume #Roll 108, 02 Jul 1862–24 Jul 1862.

Arnold, Augusta Foote. Letter to Elisha Foote. "Tables d'Olomme-Vendee July 21st 1872." Augusta Foote Arnold, 1870–1882, Smithsonian Institution Archives, Record Unit 7075, Henderson Family Papers, Box 6, Folder 2.

———. Letter to Eunice Newton Foote. "N.Y. Oct 25th 1870." Augusta Foote Arnold, 1870–1882, Smithsonian Institution Archives, Record Unit 7075, Henderson Family Papers, Box 6, Folder 2.

———. Letter to Eunice Newton Foote. "New Brighton Dec 1st 82." Augusta Foote Arnold, 1870–1882, Smithsonian Institution Archives, Record Unit 7075, Henderson Family Papers, Box 6, Folder 2.

———. Letter to Mary Foote Henderson. "N. Y. Nov. 4th 1870." Augusta Foote Arnold, 1870–1882, Smithsonian Institution Archives, Record Unit 7075, Henderson Family Papers, Box 6, Folder 2.

The Automotive Hall of Fame. "Bertha Benz." www.automotivehalloffame.org/honoree/bertha-benz.

BabyCenter. "Maud." n.d. www.babycenter.com/baby-names/details/maud-86815.

The Baltimore Sun. "The Patent Office at Washington Took Fire," September 25, 1877, p. 1. www.newspapers.com/clip/114201135/patent-office-fire-sept-1877.

Bergland, Renée. *Maria Mitchell and the Sexing of Science: An Astronomer among the American Romantics*. Boston: Beacon Press, 2008.

The Brooklyn Daily Eagle. "Rights of Married Women," February 28, 1845. www.newspapers.com/clip/109363295/1845-ny-married-women-patent-law.

Brooklyn Daily Eagle and Kings County Democrat. "Chap. 200. An Act for the More Effectual Protection of the Property of Married Women," May 20, 1848.

———. "Chap. 213. An Act Making an Appropriation to the Erie Canal Enlargement," May 20, 1848. newspaperarchive.com/brooklyn-daily-eagle-and-kings-county-democrat-may-20-1848-p-8.

Casetext. "Silsby et al. v. Foote." 61 U.S. 378 (1857), US Supreme Court. casetext.com/case/silsby-et-al-v-foote-1.

Cott, Charles M. *History of the Newton and Oviatt Families.* Columbus, Ohio: Chas. M. Cott, 1875. archive.org/details/fl-49619-tn-1394735/mode/2up.

Dick, John. "Report of the Woman's Rights Convention, held at Seneca Falls, New York, July 19th and 20th, 1848. Proceedings and Declaration of Sentiments." Rochester, New York: the *North Star* Office, July 19-20, 1848. www.loc.gov/resource/rbcmil.scrp4006702/?sp=7&st=text.

The Evansville Journal. "River Intelligence," July 27, 1871, p. 3. www.newspapers.com/clip/110402680/olive-branch-wreck-details-1871.

The Evening Star. "The Edison Light," December, 29, 1879, p. 2. www.newspapers.com/clip/114317573/edisons-light-1879.

Foote, Elisha. "Electricity in Thunder-Storms." *Popular Science Monthly* 13, October 13, 1878. en.wikisource.org/wiki/Popular_Science_Monthly/Volume_13/October_1878/Electricity_in_Thunder-Storms.

———. Improvement in Gas-Burners. US Patent 137,905, filed March 1, 1873, and issued April 15, 1873. patentimages.storage.googleapis.com/ef/68/ba/8febbe993993cb/US137905.pdf.

———. Improvement in Grain-Bands, Bag-Ties, &c. US Patent 135,899, issued February 18, 1873. patentimages.storage.googleapis.com/8b/34/84/024565d3ae27e4/US135899.pdf.

———. Improvement in Skates. US Patent 45,148, issued November 22, 1864. patentimages.storage.googleapis.com/25/d0/f7/47ebfec531ac02/US45148.pdf.

———. Letter to Green-Wood Cemetery. "Burial Orders." Green-Wood Cemetery Archives, Plot 8379, December 20, 1875.

———. Letter to Eunice Newton Foote. "Washington 11 Dec 78." Elisha Foote, 1871-1883, Smithsonian Institution Archives, Record Unit 7075, Henderson Family Papers, Box 6, Folder 17.

———. Letter to Joseph Henry. "Description of Foote's Weather Vane." National Archives and Records Administration, JHPP Reel M037, March 21, 1848.

———. Letter to Mary Foote Henderson. "Washington 11 Jan 71." Elisha Foote, 1871-1883, Smithsonian Institution Archives, Record Unit 7075, Henderson Family Papers, Box 6, Folder 17.

———. Letter to William H. Seward. "Letter Requesting Expedited Passports." Ancestry.com, July 21, 1862.

———. Machinery for Reaping and Binding Grain. US Patent 244,876, filed August 12, 1879, and issued July 26, 1881. patentimages.storage.googleapis.com/66/3a/39/5ca7541709c349/US244876.pdf.

———. Regulating the Draft of Stoves. US Patent 2,636, issued May 26, 1842. patentimages.storage.googleapis.com/fc/e7/39/80e5836b9cf31f/US2636.pdf.

———. "On the Heat in the Sun's Rays." *The London, Edinburgh, and Dublin Philosophical Magazine and Journal of Science*, 4th ser., 13, no. 85 (1857): pp. 167–172. doi.org/10.1080/14786445708642274.

Foote, Elisha, and Marshall P. Smith. Improvement in Driers. US Patent 124,944, issued March 26, 1872. patentimages.storage.googleapis.com/cf/34/dd/f25ce739e14b7b/US124944.pdf.

Foote, E. N. Filling for Soles of Boots and Shoes. US Patent 28,265, issued May 15, 1860. patentimages.storage.googleapis.com/0f/c3/8a/df35a07f37f16f/US28265.pdf.

———. Improvement in Paper-Making Machines. US Patent 45,144, issued November 22, 1864. patentimages.storage.googleapis.com/e9/3a/0b/f5f711bf530e1b/US45149.pdf.

Foote, Eunice. "Circumstances Affecting the Heat of the Sun's Rays." *The American Journal of Science and Arts*, 2nd ser., 22, no. 66 (1856): pp. 382–383. New York: G. P. Putnam & Co., 1856. archive.org/details/mobot31753002152491/page/381/mode/2up?view=theater.

Foote, Eunice Newton. Letter to Mary Foote Henderson. "Dec 7th." Eunice Newton Foote, 1870–1871, Smithsonian Institution Archives, Record Unit 7075, Henderson Family Papers, Box 6, Folder 18.

———. Letter to Mary Foote Henderson. "East Bloomfield Nov 7th." Augusta Foote Arnold, 1870–1882, Smithsonian Institution Archives, Record Unit 7075, Henderson Family Papers, Box 6, Folder 2.

———. Letter to Mary Foote Henderson. "NY Jan 1st." Eunice Newton Foote, 1870–1871, Smithsonian Institution Archives, Record Unit 7075, Henderson Family Papers, Box 6, Folder 18.

———. Letter to Mary Foote Henderson. "Washington Sunday 19th." Eunice Newton Foote, 1870–1871, Smithsonian Institution Archives, Record Unit 7075, Henderson Family Papers, Box 6, Folder 18.

Foote, Mrs. Elisha. "On a New Source of Electrical Excitation," *The London, Edinburgh, and Dublin Philosophical Magazine and Journal of Science*, 4th ser., 15, no. 99 (1858): pp. 239–240. doi.org/10.1080/14786445808642471.

Foote, Mrs. Eunice. "On a New Source of Electrical Excitation." In *Proceedings of the American Association for the Advancement of Science: Eleventh Meeting, Held at Montreal, Canada, East, August 1857*, edited by Joseph Lovering, p. 123. Cambridge: Allen and Farnham, 1858.

Foote, Henry R. Improved Vapor Generator and Burner for Heating Purposes. US Patent 63,160, issued March 19, 1867. patentimages.storage.googleapis.com/f5/4e/09/26642326b14594/US63130.pdf.

Foote, Henry Rutger. *A Description of the Apparatus for Burning Crude Petroleum in Marine and Locomotive Boilers: Invented by Col. Henry R. Foote*. New York: Baker & Godwin, 1867.

Green-Wood Cemetery Archives. Interment record, "Foote, Eunice N.," September 29, 1888, Plot 8379.

———. Interment record, "Henderson, Maud," August 28, 1872, Plot 8379.

Griffith, Elisabeth. *In Her Own Right: The Life of Elizabeth Cady Stanton*. New York: Oxford University Press, 1987.

Gurganus, Ray. "Henry Newton Arnold (1873–1939)." Find a Grave, June 2, 2014. www.findagrave.com/memorial/130771618/henry-newton-arnold.

Hamlin, Kimberly A. "Monumental Women: Adelaide Johnson, the Sculptor of Suffrage." *Humanities: The Magazine of the National Endowment for the Humanities*, August 10, 2010. www.neh.gov/article/monumental-women.

Heinemann, Sue. *Timelines of American Women's History*. New York: Berkley, 1996. tinyurl.com/EyesToSky3.

Henderson, John B. Letter to Elisha Foote. "St Louis Sept/74." John Brooks Henderson, 1868–1872, Smithsonian Institution Archives, Record Unit 7075, Box 6, Folder 21.

———. Letter to Mary Foote Henderson. "Tuesday Aug 29 4 ½ p.m." John Brooks Henderson, 1868–1872, Smithsonian Institution Archives, Record Unit 7075, Box 6, Folder 21.

———. Letter to Mary Foote Henderson. "Washington Thursday 11 p.m." John Brooks Henderson, 1868–1872, Smithsonian Institution Archives, Record Unit 7075, Box 6, Folder 21.

———. Letter to Mary Foote Henderson. "Washington July 17 1872." John Brooks Henderson, 1868–1872, Smithsonian Institution Archives, Record Unit 7075, Box 6, Folder 21.

———. Letters to Mary Foote Henderson. "Monday 13 Aug.," "St. Louis Mo Saturday 17," and "Monday morning Aug 19." John Brooks Henderson, 1868–1872, Smithsonian Institution Archives, Record Unit 7075, Box 6, Folder 21.

Henderson, Mrs. Mary F. *Practical Cooking and Dinner Giving: A Treatise Containing Practical Instructions in Cooking; in the Combination and Serving of Dishes; and in the Fashionable Modes of Entertaining at Breakfast, Lunch, and Dinner*. New York: Harper & Brothers, 1876. d.lib.msu.edu/fa/59#page/246/mode/2up.

Henderson, Mary Foote. *Diet for the Sick: A Treatise on the Values of Foods, Their Application to Special Conditions of Health and Disease, and on the Best Methods of Their Preparation*. New York: Harper & Brothers, 1885.

———. Letter to May Wright Sewall. "February 16, 1890." Indianapolis Special Collections Room, Indianapolis Public Library, May Wright Sewall Letters. Digitized by Indianapolis-Marion County Public Library, 2005.

———. Letter to Mother (Eunice Newton Foote). "St. Louis 28th Sept. '72." Mary Foote Henderson, 1872–circa 1880, Smithsonian Institution Archives, Record Unit 7075, Henderson Family Papers, Box 7, Folder 6.

———. Letter to Mother (Eunice Newton Foote), Father (Elisha Foote) & Guss (Augusta Foote Arnold). "Sunday." Mary Foote Henderson, 1872–circa 1880, Smithsonian Institution Archives, Record Unit 7075, Henderson Family Papers, Box 7, Folder 6.

———. Letter to Mother, Father & Guss. "Idaho Springs Monday." Mary Foote Henderson, 1872–circa 1880, Smithsonian Institution Archives, Record Unit 7075, Henderson Family Papers, Box 7, Folder 6.

———. Letter to Mother, Father & Guss. "Idaho Springs Col Sunday Aug. 1874." Mary Foote Henderson, 1872–circa 1880, Smithsonian Institution Archives, Record Unit 7075, Henderson Family Papers, Box 7, Folder 6.

Henry, Joseph. *The Scientific Writings of Joseph Henry*. Washington, DC: Smithsonian Institution, 1886.

HerrGesell, Leif. Email to author, June 1, 2023.

Hume, Robert. "John Tyndall: Remembering the Carlow Man Who Discovered Why the Sky Is Blue." *Irish Examiner*, August 7, 2020. www.irishexaminer.com/lifestyle/people/arid-40028360.html.

Ithaca Daily Chronicle. "Unrivaled: Foot's [sic] Patent Self Regulating Stove," September 28, 1847.

Justia. "Silsby v. Foote, 55 U.S. 218 (1852)." US Supreme Court, 1852. supreme.justia.com/cases/federal/us/55/218.

Leavenworth Daily Conservative. "An Improved Cylinder Paper Machine," November 16, 1867. www.newspapers.com/clip/89786975/eunice-newton-foote-cylinder-paper.

Lewis, Jone Johnson. "Seneca Falls Resolutions: Women's Rights Demands in 1848." ThoughtCo, last modified July 3, 2019. www.thoughtco.com/seneca-falls-resolutions-3530486.

Library of Congress. "Our Roll of Honor. Listing Women and Men Who Signed the Declaration of Sentiments at First Woman's Rights Convention, July 19–20, 1848." hdl.loc.gov/loc.rbc/rbcmil.scrp4006701.

McCully, Emily Arnold. "The Oil Region." In *Ida M. Tarbell: The Woman Who Challenged Big Business—and Won!* pp. 7–9. New York: Clarion Books, 2014.

McKenzie, Samuel. Email to author, June 2, 2023.

Merriam-Webster Dictionary. "Carbon dioxide." www.merriam-webster.com/dictionary/carbon%20dioxide.

National Park Service. "Declaration of Sentiments." Women's Rights National Historical Park, last modified February 7, 2023. www.nps.gov/wori/learn/historyculture/declaration-of-sentiments.htm.

———. "Participants of the First Women's Rights Convention." Women's Rights Historical Park, last modified April 4, 2023. www.nps.gov/wori/learn/historyculture/participants-of-the-first-womens-rights-convention.htm.

New York Daily Herald. "Fashionable Intelligence: The Marriage of Senator Henderson in Washington—a Brilliant Affair," June 27, 1868. www.newspapers.com/clip/95163209/mary-henderson-wedding-full-details-1868.

———. "The Scientific Congress: American Association for the Advancement of Science," August 26, 1856. www.newspapers.com/clip/117172132/aaas-mtg-in-albany-august-1856.

New York Daily Times. "The Scientific Convention, Third Day," August 18, 1857. timesmachine.nytimes.com/timesmachine/1857/08/18/issue.html.

New York Department of State. "New York State Emblems." dos.ny.gov/new-york-state-emblems.

The New York Times. "East and West: Completion of the Great Line Spanning the Continent," May 11, 1869. www.newspapers.com/clip/11689/the-new-york-times.

The North Star. "The Rights of Women," July 28, 1848, p. 3.

The Nunda (New York) Democrat. "Regulator Ahead!" 1848. tinyurl.com/EyesToSky.

Ortiz, Joseph D., and Roland Jackson. "Understanding Eunice Foote's 1856 Experiments: Heat Absorption by Atmospheric Gases." *Notes and Records: The Royal Society Journal of the History of Science* online, August 26, 2020. doi.org/10.1098/rsnr.2020.0031.

The Philadelphia Inquirer. "A National Calamity," September 25, 1877, p. 1. www.newspapers.com/clip/114201338/patent-office-fire-sept-1877.

Proceedings of the American Association for the Advancement of Science: First Meeting, Held at Philadelphia, September, 1848. Philadelphia: John C. Clark, 1849.

Race, Washburn. Improvement in Registers for Stoves. US Patent 4,443, issued April 4, 1846. patentimages.storage.googleapis.com/ac/be/16/ebe36191cc790c/USRE1139.pdf.

The Revolution, masthead. archive.org/details/revolution-1868-04-16.

Ricketts, Palmer C. *History of the Rensselaer Polytechnic Institute, 1824–1894*. New York: John Wiley and Sons, 1895. archive.org/details/cu31924021892777/page/n95/mode/2up.

Riddick, Floyd M., and Robert B. Dove. *Procedure and Guidelines for Impeachment Trials in the United States Senate*. Washington, DC: United States Senate, 1986.

Robb, Frances C. "Erie Canal." *Britannica*, last modified March 8, 2018. www.britannica.com/topic/Erie-Canal.

Robertson, Charles J. *Temple of Invention: History of a National Landmark*. Washington, DC: Smithsonian American Art Museum National Portrait Gallery; London: Scala Publishers, 2006.

Ross, Sydney. "Scientist: The Story of a Word." *Annals of Science* 18, no. 2 (1964). doi.org/10.1080/00033796200202722.

Saratoga. "Elisha Foote (1809-1883)." Find a Grave, May 29, 2012. www.findagrave.com/memorial/90957196/elisha-foote.

Search and Discovery. "About Search and Discovery." AAPG, Datapages, n.d. www.searchanddiscovery.com/about.html.

Seneca County Courier. "Stoves! Stoves!" Oct. 30, 1844. tinyurl.com/EyesToSky12.

Smithsonian Institution Archives. "James Smithson, Founding Donor," March 9, 2011. www.siarchives.si.edu/history/james-smithson.

The Southern Argus. "Mrs. Willard." September 17, 1839, p. 3. newspaperarchive.com/southern-argus-sep-17-1839-p-3.

Stanton, Elizabeth Cady. "Washington." *The Revolution*, April 16, 1868. archive.org/details/revolution-1868-04-16/page/n1.

St. Louis Post-Dispatch. "Hon. Elisha Foote: Death of the Father-in-Law of Gen. Henderson This Morning," October 22, 1883, p. 5. www.newspapers.com/clip/89828430/elisha-death-longer-description.

Tennyson, Alfred, Lord. "From Maud (Part XVIII)." Poetry Foundation, n.d. www.poetryfoundation.org/poems/50286/maud-part-xviii-i-have-led-her-home-my-love-my-only-friend.

Trickey, Erick. "'Kill the Beast': The Impeachment Trial That Nearly Took Down a President 150 Years Ago." *Washington Post*, May 16, 2018. www.washingtonpost.com/news/retropolis/wp/2018/05/16/kill-the-beast-the-impeachment-trial-that-nearly-took-down-a-president.

Tyndall, John. "VII. Note on the Transmission of Radiant Heat through Gaseous Bodies." *Proceedings of the Royal Society of London* 10 (December 1860): pp. 37–39. doi.org/10.1098/rspl.1859.0017.

United States Senate. "The Civil War: The Senate's Story," n.d. www.senate.gov/artandhistory/history/common/civil_war/LincolnEmergencySession_FeaturedDoc.htm.

US Patent and Trademark Office. "Elisha Foote, 1868–1869." n.d. www.uspto.gov/about-us/elisha-foote.

Wellcome Collection. *Proceedings of the Third Race Betterment Conference, January 2-6, 1928: Under the Auspices of the Race Betterment Foundation, Battle Creek, Michigan*. Battle Creek, MI: Race Betterment Foundation, 1928. wellcomecollection.org/works/ex9tb25e.

Wells, David A., ed. *The Annual of Scientific Discovery: or Year-book of Facts in Science and Art: For [...] Exhibiting the Most Important Discoveries and Improvements in Mechanics, [...] 1857*. Boston: Gould and Lincoln, 1857.

Willard, Emma. *An Address to the Public; Particularly to the Members of the Legislature of New York, Proposing a Plan for Improving Female Education*. Middlebury, VT: J. W. Copeland, 1819. babel.hathitrust.org/cgi/pt?id=mdp.39015018046972&seq=11.

Selected Additional Bibliography

I consulted more than 250 sources—too many to list. The following are the most pertinent sources not quoted from directly.

Primary Sources

Adams, John Quincy. "Rochester Friday 28 July 1843 Canandaigua, Auburn." *The Diaries of John Quincy Adams: A Digital Collection*. Massachusetts Historical Society, 2023. www.masshist.org/jqadiaries/php.

Albany Academy. "Foote, Elisha, Jr." Academic record, 1827–1830.

Albany Express. "Washington Affairs," August 1, 1868. tinyurl.com/EyesToSky4.

Albany Morning Express. "Field Crops," February 10, 1858. tinyurl.com/EyesToSky6.

Ancestry.com. "Connecticut, U.S., Town Birth Records, pre-1870 (Barbour Collection)." Lehi, UT: 2006. Original data: White, Lorraine Cook, ed. *The Barbour Collection of Connecticut Town Vital Records*. Vol. 1–55. Baltimore: Genealogical Publishing, 1994–2002.

———. "The Foote family: or, The Descendants of Nathaniel Foote, One of the First Settlers of Wethersfield. North America, Family Histories, 1500–2000." Provo, UT, 2016.

———. "Isaac Jr., his w., dism. & Recon. Sept. 19, 1824," "Connecticut, U.S., Church Record Abstracts, 1630–1920." Provo, UT, 2013. Original data: Connecticut, Church Records Index. Connecticut State Library, Hartford, Connecticut.

Buffalo Morning Express. "Circuit Court of the United States," August 29, 1849. www.newspapers.com/clip/110475107/buffalo-morning-express.

Bull, Mary A. "Woman's Rights and Other 'Reforms' in Seneca Falls." *Seneca Falls Reveille*, July 9, 1880.

Case Law Access Project, Harvard Law School. "Foote v. Silsby, 9 F. Cas. 391, 3 Blatchf. 507 (1856)." United States Circuit Court for the Northern District of New York, August 28, 1856. cite.case.law/f-cas/9/391.

———. "Foote v. Silsby, 9 F. Cas. 373, 1 Blatchf. 445, 1 Fish. Pat. Rep. 268 (1849)." United States Circuit Court for the Northern District of New York, October 1849. cite.case.law/f-cas/9/373/11644092.

Charleston Courier. "Arrival of the Congressional Junketing Party—the Charlestonians Stared At—Savannah to Be Viewed Next—List of the Distinguished Visitors," March 20, 1865. tinyurl.com/EyesToSky5.

The Daily Messenger. "Aged Cleveland Man Gives $100,000 to Bristol Churches," April 27, 1925. www.newspapers.com/clip/88778327/newton-gives-100000-page-1.

Durkee, Cornelius E., comp. *Reminiscences of Saratoga*. Reprinted from *The Saratogian*, 1927–28, p. 182.

Fairbanks, Mary J. Mason (Mrs. Abel W.), ed. *Emma Willard and her Pupils or Fifty Years of Troy Female Seminary 1822–1872*, pp. 213–214. New York: Mrs. Russell Sage, 1898. archive.org/details/cu31924030634921/page/5/mode/1up.

Green-Mountain Freeman. "The Mercantile Journal Says," December 10, 1842. www.newspapers.com/clip/110476822/foote-self-regulating-stove-10-dec-1842.

Henry, Joseph. Letter to J. D. Cox, Secretary of the Interior. "Recommendation for Elisha Foote." National Archives and Records Administration, RG 48 X164 34437, March 25, 1869.

———. Letter to J. Q. Gillmore. "Henry Calls Elisha a Friend." Smithsonian Institution Archives, Record Unit 33, Office of the Secretary, Outgoing Correspondence, M316, June 21, 1866.

Holly, Birdsall. *Condensing Steam-Engines Which Are Used for Pumping*. US Patent 14,239, issued February 12, 1856. patentimages.storage.googleapis.com/95/f7/ff/dc611682c26dbe/US14239.pdf.

Lee, Dick. "Nurse Denies Dowager's Waif Story." *Daily News*, February 5, 1931. www.newspapers.com/clip/95176029/mary-henderson-john-quincy-adams.

Leonard, Ermina Newton, comp. *Newton Genealogy, Genealogical, Biographical, Historical: Being a Record* [...] *Near Boston*. De Pere, WI: Bernard Ammidown Leonard, 1915. tinyurl.com/EyesToSky14.

The Lily. "Woman's Mechanical and Manual Powers," July 1, 1851.

New York Daily Herald. "Married," March 8, 1869, p. 9. www.newspapers.com/clip/105267035/new-york-daily-herald.

New York Daily Tribune. "Appointments by the Governor," March 16, 1846. www.newspaperarchive.com/politics-clipping-mar-16-1846-3306016.

The New York Times. "The Foote Case," January 15, 1865. timesmachine.nytimes.com/timesmachine/1865/01/15/issue.html.

———. "Patent Office Appointment," August 1, 1865. newspaperarchive.com/politics-clipping-aug-01-1865-3308039.

———. "Petroleum as Fuel," January 22, 1868. timesmachine.nytimes.com/timesmachine/1868/01/22/78906046.pdf.

New York (State) Court of Chancery. *Reports of Cases Argued and Determined in the Court of Chancery of the State of New York* [...], vol. 3. New York: Banks & Brothers, 1887.

Ontario County Journal. "Mrs. Amanda Newton of East Bloomfield," November 24, 1882. tinyurl.com/EyesToSky7.

Palmer, Jefferson. *Manuscript diary, July 1848*. Montezuma, New York: Seneca Falls Historical Society, n.d.

The Saratogian. "Invention by Mrs. Judge Foote." Vol. 16, no. 12, March 12, 1868, p. 3. tinyurl.com/EyestoSky9.

Scientific American. "Patent Cases in Court." Vol. 20, no. 18 (May 1, 1869), p. 283. doi.org/10.1038/scientificamerican05011869-283a.

——— "Scientific Ladies.—Experiments with Condensed Gases." Vol. 12, no. 1 (September 13, 1856), p. 5. babel.hathitrust.org/cgi/pt?id=coo.31924080787660&view=1up&seq=9&size=175&q1=sun%20lady.

Seneca Falls Historical Society. "Map of Seneca Falls in 1856." Includes the notation, "4 North Park Street, E. Foot."

Smithsonian Institution. *Meteorological Stations and Observers of the Smithsonian Institution in North America and Adjacent Islands, from the Year 1849 up to the Year 1869*. Washington, DC: Smithsonian Institution, n.d.

Vermont Phoenix. "Another Divorce Case in High Life," September 6, 1839. newspaperarchive.com/brattleboro-vermont-phonix-sep-06-1839-p-1.

Washington National Intelligencer. "Troy Female Seminary," January 24, 1839, p. 1. newspaperarchive.com/washington-national-intelligencer-jan-24-1839-p-1.

Youmans, Eliza Ann. "Sketch of Edward L. Youmans." *Popular Science Monthly* 10 (March 1887). en.wikisource.org/wiki/Popular_Science_Monthly/Volume_30/March_1887/Sketch_of_Edward_L._Youmans.

Secondary Sources

Conway, W. Fred. *Those Magnificent Old Steam Fire Engines.* New Albany, IN: Fire Buff House Publishers, 1997.

Dobyns, Kenneth W. *The Patent Office Pony: A History of the Early Patent Office.* Fredericksburg, VA: Sergeant Kirkland's Museum and Historical Society, 1994.

Hunt, Bill. "Signer #5, Eunice Newton Foote: 'the Climate Scientist.'" 100 Signers Project, June 2020. www.100signersproject.com/signer-profiles/signer-5-eunice-newton-foote-the-climate-scientist.

Jackson, Roland. "Eunice Foote, John Tyndall and a Question of Priority." *Notes and Records: The Royal Society Journal of the History of Science* 74 (February 13, 2019), pp. 105–118. doi.org/10.1098/rsnr.2018.0066.

Khan, B. Zorina. *The Democratization of Invention: Patents and Copyrights in American Economic Development, 1790–1920.* Cambridge, England: Cambridge University Press, 2005.

McKenzie, Samuel. "Eunice Newton Foote and Her Family at Saratoga Springs, 1859–1865," unpublished manuscript. Saratoga Springs Public Library.

Perlin, John. "Symposium: Science Knows No Gender? Eunice Foote." Renewables 100 Policy Institute, May 17, 2017. www.youtube.com/watch?v=yHc5Ike2GIA.

Seneca Museum of Waterways and Industry. "Double Weathered Clapboard Being Painted Yellow Built Before 1846." 1848 Seneca Falls Diorama Exhibit.

Stanton, Elizabeth Cady, Ann Dexter Gordon, Susan Brownell Anthony, Tamara Gaskell Miller, and Stacy Kinlock Sewell. *The Selected Papers of Elizabeth Cady Stanton and Susan B. Anthony: In the School of Anti-Slavery, 1840 to 1866.* New Brunswick, NJ: Rutgers University Press, 1997.

Stanton House Women's Rights National Historical Park. "National Park Service Cultural Landscapes Inventory," 1998, pp. 21–24. irma.nps.gov/DataStore/DownloadFile/454362.

Weber, Sandra S. "Special History Study: Women's Rights National Historical Park, Seneca Falls, New York." US Department of the Interior / National Park Service, 1985. www.npshistory.com/publications/wori/shs.pdf.

Wellman, Judith. *The Road to Seneca Falls: Elizabeth Cady Stanton and the First Woman's Rights Convention.* Urbana, IL: University of Illinois Press, 2004.

Image Credits

Title page, ii–iii: photograph and outline by Sarah Coleman.

Part openers, pp. viii–ix; pp. 32–33; pp. 164–165: AdobeStock, image #648917115 by Hamza.

p. 152: Lindsay H. Metcalf.

p. 153 (top): *Emma Willard. The Life of Emma Willard by John Lord*, 1873. Harvard Library. iiif.lib.harvard.edu/manifests/view/drs:2573577$1i.

p. 153 (bottom): Troy Female Seminary science lab photo, n.d., Emma Willard School Archives, Box 5, Troy, NY.

p. 154 (top): Ancestry.com, *Eunice N. Foote*, U.S. Passport Applications, 1795–1925.

p. 154 (bottom): Wikimedia Commons, photograph of Judge Elisha Foote.

p. 155 (top left): Library of Congress, Prints and Photographs Division (LCCN: 93512850).

p. 155 (top right): Library of Congress, Prints and Photographs Division (LCCN: 2017894687).

p. 155 (bottom): Wikimedia Commons, photograph of Augusta Foote Arnold.

p. 156 & 157: Wikimedia Commons, *Dudley Observatory Dedication, August 28, 1856*, Tompkins Harrison Matteson, 1857.

p. 158 (top): Google Patents, *Filling for Soles of Boots and Shoes*, Eunice N. Foote, patent No. 28,265, 1860.

p. 158 (bottom): Google Patents, *Paper Making Machine*, Eunice N. Foote, patent No. 45,144, 1864.

p. 159 (top): Google Patents, *Machinery for Reaping and Binding Grain*, Elisha Foote, patent No. 244,876, 1881.

p. 159 (bottom): Smithsonian Institution Archives (ID: 2003-19427).

p. 160 & 161: *The American Journal of Science and Arts*, Vol. XXII, No. 65, September 1856, pp. 382–383. archive.org/details/mobot31753002152491/page/381

p. 162 (top): Library of Congress, Prints and Photographs Division (LCCN: 2006682742).

p. 162 (bottom): Library of Congress, Prints and Photographs Division (LCCN: 2006683456).

p. 163 (top): Library of Congress Prints, and Photographs Division (LCCN: 2017892970).

p. 163 (bottom): Library of Congress, The Alfred Whital Stern Collection of Lincolniana (loc.gov/item/scsm000882).

p. 268: Lindsay H. Metcalf, Detail of *Dudley Observatory Dedication, August 28, 1856*, Tompkins Harrison Matteson, 1857. Albany Institute of History & Art.

Acknowledgments

Thank you to Sandra Yeaton, a great-great-granddaughter of Eunice Newton Foote, for answering the phone and graciously scouring family memorabilia in search of a portrait of Eunice. Thanks to Lurie Emmons, whose father, Jesse Shima, worked for Mary Foote Henderson and became heir to her estate. Lurie also joined in the search for a portrait of Eunice.

My gratitude to all the archivists, librarians, curators, and researchers who helped me locate documents and information: Tad Bennicoff of the Smithsonian Institution Archives; Christina Bickford, Connie Thatcher, and Jeff Richman of the Green-Wood Cemetery genealogy team; Stephanie Ross of the Emma Willard School; Leif R. HerrGesell, Ontario County (NY) historian; Barbara Hunt of the East Bloomfield (NY) Historical Society; Daisy Nicosia of the Seneca Falls Historical Society; John McClintock of Albany Academy; Brooke Morse of Ontario County (NY) Records and Archives; researcher John Perlin; Brendan Aquilino of the American Association for the Advancement of Science; Jane Rogers, sports collection curator for the National Museum of American History; Diane Shewchuk of the Albany Institute of History & Art; Allison Mills of special collections at Bryn Mawr College; Michelle Isopo of the Saratoga Springs Public Library; and Samuel McKenzie, formerly of the Saratoga County History Center.

Thanks to the staff at the Frank Carlson Library, my home base for research, especially Donna Reedy for hunting down obscure titles via interlibrary loan.

To Sarah Coleman, whose cover art took my breath away.

And to my editor, Karen Boss, whose enthusiasm for Eunice's story almost matches my own, and to the whole wonderful team at Charlesbridge: designer Diane Earley, copyeditor Andie Divelbiss, and proofreader Mira Singer.

To my agent, Emily Mitchell, who helped me find the perfect home for this story—thank you.

To the beloved, generous, and encouraging critique partners who read parts or all of this story: Ali Bovis, Meeg Pincus, Angela Burke Kunkel, Candy Wellins, M. O. Yuksel, Julie Rowan-Zoch, and Vicky Fang . . . to Keila Dawson, who mined Ancestry.com records for Eunice's Newton lineage and relationship to Sir Isaac Newton . . . to Cordelia Jensen for her expertise in novels-in-verse and her courses through the Highlights Foundation . . . and to my fellow NIV classmates who offered kind words of encouragement as I revised, I'm forever indebted.

Not least, I thank my husband, Will, for supporting me all the times I needed to disappear to work on this project. And my sons, Quinn and Bennett, who tromped all over Washington, DC, and upstate New York with me in the name of research.

This book is clearly a community effort. It's because of all of you that I've been able to piece together Eunice's story as completely as I have. I hope I've done her justice.